Dedication

For my son Oliver because I am
borrowing this earth from him.

And for everyone who does their best
to make this world a better place (except for
the ones who litter, you all can eat it).

Quarto
Knows

Inspiring | Educating | Creating | Entertaining

Brimming with creative inspiration, how-to projects, and useful information to enrich your everyday life, Quarto Knows is a favorite destination for those pursuing their interests and passions. Visit our site and dig deeper with our books into your area of interest: Quarto Creates, Quarto Cooks, Quarto Homes, Quarto Lives, Quarto Drives, Quarto Explores, Quarto Gifts, or Quarto Kids.

Text © 2020 by Megean Weldon

First published in 2020 by Rock Point,
an imprint of The Quarto Group,
142 West 36th Street, 4th Floor,
New York, NY 10018, USA
T (212) 779-4972 F (212) 779-6058
www.QuartoKnows.com

Rock Point titles are also available at discount for retail, wholesale, promotional, and bulk purchase. For details, contact the Special Sales Manager by email at specialsales@quarto.com or by mail at The Quarto Group, Attn: Special Sales Manager, 100 Cummings Center Suite 265D, Beverly, MA 01915 USA.

10 9 8 7 6 5 4 3 2 1

ISBN: 978-1-63106-658-0

Library of Congress information available upon request.

Publisher: Rage Kindelsperger
Creative Director: Laura Drew
Managing Editor: Cara Donaldson
Project Editor: Keyla Pizarro-Hernández
Cover and Interior Design: Evelin Kasikov

Printed in Singapore

AN ALMOST

ZERO WASTE LIFE

LEARNING HOW TO
EMBRACE LESS TO LIVE MORE

MEGEAN WELDON

ROCK
POINT

INTRODUCTION

In the not so distant past, on a day that I now refer to as Earth Day First, I had no idea that a single day's events would rattle me to my core and forever change my life.

⬳⬳⬳⬳⬳⬳⬳⬳

What originally began as a love/hate relationship with trash ultimately led me to examine and eliminate all the waste in my life that I possibly could do without. It is now somewhat difficult to imagine my life and daily routines before this fateful day because I now live a much simpler life—full of purpose, intention, adventure, and happiness. I've even saved money doing so.

Since then, I've been on a mission to explore new ways to live without disposable waste and seek out new sustainable habits and ideas. I've found ways to avoid generating waste through preparation and by curbing my consumption to eventually eliminate most of my waste. I'm not suggesting that you NEED to cut down waste to this extreme, but I'm sharing my story to help you live a simpler, cleaner, and less stressful lifestyle.

I was born and raised in rural Northwest Missouri where my closest neighbor was five miles away, which gave me plenty of open space to simply be outdoors. I learned to love and respect the earth from an early age and was interested in everything involving nature and the world around me.

Nature has always been a big part of my life and has helped shaped who I am today. I was fortunate to live in a family that valued time outdoors, and my passion for nature followed me into adulthood.

Fast-forward a couple of decades, and I'm now an IT manager with an active three-year-old son who is as crazy for nature as I am (I can tell by the dirt rings he leaves in the tub). I've always considered myself an environmentalist with a love for the "green lifestyle."

Because of this, I've always celebrated Earth Day. The ways I showed my appreciation for the earth on this day included recycling, planting fruit trees in my backyard, and starting a garden. On an Earth Day five years ago, which I call my Earth Day First, I decided to pick up waste around my neighborhood. It was after a long workday, and I was tired, so it seemed like an easy way to honor our planet without reinventing the wheel.

I got my dog, put on her leash, grabbed a trash bag, and headed out on what I thought would be a quick trek around the block. I wasn't expecting to pick up much trash in half a mile. Boy, was I wrong.

I filled the first bag within minutes. Then, I walked back to the house to get more trash bags.

Five or so bags later, I was fuming. How could the people in my community be so careless? How could they be so wasteful? I was furious until I went to open my own trash can in front of my house to dump the bags. It was completely full, and it wasn't trash day until the end of the week. In that moment, I realized that I was as wasteful as these complete strangers that had made me so angry. I called myself an environmentalist but wasn't I just as wasteful? I felt like a hypocrite.

Until this pivotal moment, I didn't realize how my normal, everyday habits were hurting the planet. But now I knew I needed to make some changes, and fast. So, I researched (and researched!) and stumbled upon a few helpful tips, but nothing life changing. I quickly realized I had to get creative and use common sense to reduce my trash for good.

I started by replacing disposables like bags and water bottles with burlap and glass. I also learned that I could stretch the food in my pantry and refrigerator. It was a start, and I was no longer mad at myself for not knowing better. Transformations don't happen overnight, and this was my beginning.

This book is about so much more than trash.

IT'S ABOUT SAVING MONEY.

IT'S ABOUT SIMPLIFYING YOUR LIFE.

IT'S ABOUT REALISTIC ALTERNATIVES AND HABITS.

IT'S ABOUT LETTING GO OF WHAT YOU THOUGHT YOU KNEW.

IT'S ABOUT GETTING HEALTHIER, BOTH PHYSICALLY AND MENTALLY.

Living without waste isn't a new idea. People have been doing it for generations. Living mindfully is a gift from our beloved ancestors, countless indigenous peoples whose cultures and ways of life depended on frugality, and others who had no choice but to live this way because they couldn't afford to waste a single cent.

I'm here to guide you, providing alternatives to reducing waste. Just remember: This journey isn't about perfection. We all have different stories, potential, and limitations—and that's okay!

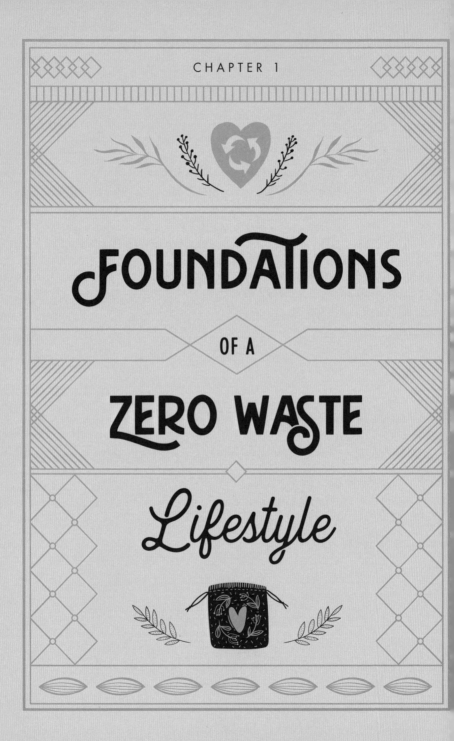

CHAPTER 1

FOUNDATIONS

OF A

ZERO WASTE

Lifestyle

Zero waste living may seem impossible, overwhelming, and downright insane, and to some degree, it is (I'm not going to lie).

‹‹‹‹‹‹‹‹‹‹‹‹

But zero waste living isn't about creating absolutely ZERO trash—it's about setting realistic expectations because in our world of convenience and instant gratification, living with no impact on our environment is impossible.

For me, it's about creating as little waste as you can in a healthy way (think "low-waste" lifestyle). I still like the term "zero waste" because that's our ultimate goal, and although you'll never get there, it's good to challenge yourself to see how close to zero you can get in a way that is sustainable to you and your family.

Each habit should be realistic and doable. Zero trash living will hopefully add a lot of value to your life and simplify an otherwise chaotic world. You will overhaul nearly every area of your life in ways you never knew needed attention and change your view on consumption.

There's this assumption that once you commit to going zero waste you must purge all plastic items from your house and purchase sustainable, nonplastic counterparts. I'm here to tell you that's NOT true. If you purged all the perfectly useful items from your life to buy new, that would negate the reasoning behind zero waste in the first place.

THIS WHOLE LIFESTYLE WILL CONNECT YOU BACK TO THE EARTH AND WHAT'S TRULY IMPORTANT.

And while the goal isn't to live a perfectly plastic-free life, try to avoid plastic whenever you can because plastic has a limited number of recycling life cycles. Once plastics' life cycles have been completely maxed out, they are almost always sent to the landfill. Metal and glass containers can also be easily upcycled, so I tend to use jars for gifts and food storage.

Now you're probably wondering what to do with the current list of plastic and disposables in your house. As tempting as it may be to toss your single-use disposables, I suggest using them first. Then, if you find items that you won't use again, donate them to local churches, schools, daycares, or nursing homes. This will help you take ownership of the items you've already purchased, seeing as there's no sense in creating more waste in the attempt to reduce waste. As a cultivated consumer, you shouldn't feel bad that you have these items in your home. We all do. And remember, zero waste living isn't about being perfect.

Now that you've committed to not purchasing these disposable or plastic items anymore, what should you do with your existing plastic collection? Here are some suggestions.

Use It

Use it until it wears out—as long as it doesn't compromise your health. Aside from cooking in your plastic bowls (please, don't do this!), use and wear out those containers, toothbrushes, hairbrushes, and razors. Once these types of plastic items are worn out, replace them with the eco-friendly options.

We all have plastic items in our homes that are perfectly usable. Living "zero waste" means making better, more sustainable choices going forward and utilizing what we already have. It's not about buying more, but about buying less.

Okay, so you're not throwing anything away for the sake of reducing clutter. Good! Now what?

ZERO WASTE IS ABOUT BEING BETTER WHEN YOU CAN AND ENACTING SMALL CHANGES THAT MAKE A BIG DIFFERENCE.

Assess Your Mess

In order to identify what to keep, you need to know what you're throwing away in the first place.

Keep a tally or journal of the trash you make in a week's time. Notice where your biggest opportunities are to get a high-level view of your trash habits.

By auditing your trash, you'll hopefully learn a few things:

- How much trash do you produce?
- What category does most of your waste fall under?
  - » Food Packaging
  - » Organic Waste
  - » Take-Out
  - » Other
- Could any of the items you throw away be easily made?

You can get a good idea of your trash health from a week's audit, but feel free to do it as long as you want. At this point, you're gathering data to get a better understanding of where to put more focus when it comes to your waste reduction.

After you've gotten a better idea of what you're throwing away, dump the idea that living a more sustainable life has everything to do with buying a bunch of stuff. The goal here is to make waste reduction as achievable as possible

so you'll be more inclined to use what you already own. Of course, there will be things you'll want to purchase. These will be sustainable alternatives that can't be easily replaced with something you already own, like a toothbrush for example.

What are some of the items in your home that can help reduce waste? Take a look at this top ten list:

❶ Reusable Bags

These are easy to acquire, considering many businesses use them as a promotional item. If you don't have reusable cloth bags, there are tutorials online that will show you how to turn an unwanted T-shirt into a grocery bag. You can also ask friends and family if they have any to spare. Avoid purchasing any unless it's necessary.

2 Reusable Water Bottles

Use a reusable glass water bottle instead of the plastic variety. You can even use a mason jar or an empty pasta sauce jar to hold water or other beverages.

3 Paper Towel Alternatives

Ripped-up old towels, shirts, sheets, curtains, or socks work just fine to dust or clean surfaces. It's not about being fancy, it's about being resourceful.

4 Reusable Coffee Cup

Put your java in a jar. If you're worried about the glass being too hot, wrap a sock, mitten, or headband around the jar for easy handling.

5 Cloth Produce Bags

Any drawstring, cloth bag you own will work for this. If you want to get crafty, you could make a few bags yourself or have a sewing day with a friend to make a set out of old sheets, curtains, or any scrap fabric.

6 Reusable Utensils

Spoons and forks right out of your utensil drawer are better than disposables. Just stick a fork in your backpack or purse and you're all set. If you don't want to use your kitchen cutlery, use the plastic ones you have in your kitchen.

7 To-Go Containers

Those stainless steel containers with pretty patterns are expensive, and chances are, you already own something that will work just as well. Once again, mason jars can be used for transporting food on the go as well as the plastic storage containers you already have in your cabinets.

8 Straws

If you don't want to buy reusable straws, just stop using plastic straws entirely. You don't always need to find a replacement to reduce your waste.

9 Reusable Tissues and Hankies

This is another great job for the DIY rags from above. Instead of using countless tissues and flushing them down the toilet, keep a pretty, handmade handkerchief on you and wash as needed.

10 Plastic Wrap Alternatives

Instead of using plastic wrap, put a plate on top of a bowl to cover it. Easy, right? You can also put a damp towel over a bowl to hold in moisture if you're storing something like rice or dough.

Since I'm most tempted to waste while I'm on the go, I keep a "zero waste kit" on me at all times that includes the following items:

- Bag, backpack, or purse that contains reusable items
- Cloth napkins or homemade rag to avoid paper napkins
- Utensils
- Reusable water bottle or dual-purpose jar (perfect to store leftovers at a restaurant!)

Now that you have a general idea of how this zero waste lifestyle will work, let's start with the biggest trash producer in the house: your kitchen.

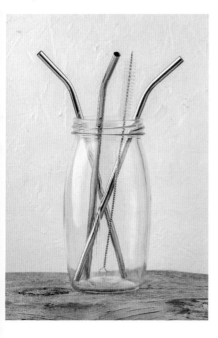

CHAPTER 2

KITCHEN, FOOD WASTE,

AND

COMPOSTING

Most of your waste comes from the kitchen.
It is the home of your garbage can, is it not?

─‹‹‹‹‹‹‹‹‹‹‹─

In this chapter, we'll tackle the single-use disposables and save a whole boatload of money in the process of finding better alternatives. There is quite a lot of expense in replacing the disposables that we use in our kitchens on a weekly or monthly basis.

I find it easier to dissect spaces based off the disposables that reside in them. It's important that you assess each item you throw away and ask yourself some important questions, like: "Do I really need this thing?" or "Do I have something that will do the job already?" I'm confident that most of the things you have in your home will do nearly any job that you use disposables for.

On the next page you'll see a list of the disposables that currently take up space in your kitchen (you will use this list throughout this chapter as a checklist of what you've stopped using).

TACKLING THE WASTE IN THE KITCHEN WILL BE THE MOST REWARDING PART OF THIS JOURNEY. BUT IT WILL ALSO BE THE HARDEST.

DISPOSABLES CHECKLIST:

- [ ] Paper Towels
- [ ] Coffee Filters
- [ ] Paper Napkins
- [ ] Disposable Plates
- [ ] Disposable Cutlery
- [ ] Plastic Wrap
- [ ] Sandwich Bags
- [ ] Trash Bags
- [ ] Tea Bags
- [ ] Coffee Pods
- [ ] Dish Sponges
- [ ] Dish Soap

USE YOUR DISPOSABLES

When using disposables, try to keep them out of the landfill once you're done. Recycle them if you can. For paper products, recycle or compost them at the end of their life. Remember, trash is only trash if you throw it away. It's impossible to ensure everything is disposed of properly, so if most items end up in the trash, just remember that it's okay and that it's part of the process.

And who's to say you must throw disposables away? You can use foil and zipper bags more than once. You can use a balled-up piece of aluminum foil to scrape off stuck-on food from your grill or a gallon zipper bag to store frozen bananas. And as those disposables disappear, we will work on zero waste solutions to do the job going forward. Here are some disposables you can eventually do without.

PAPER TOWELS

Countless resources go into producing paper towels over and over. By cutting your first disposable from your shopping list, you now have a little extra cash for more important things, like healthier food, experiences, or your bank account. Imagine that! Paper towel replacement? Swap them for rags from old clothing or scrap fabric.

How many single socks do you have lying around? Rather than let them sit in the laundry room, put them to work. Another option is to find some old, ratty clothes and cut them up. Aside from socks and shirts, other types of clothing

like flannels and jeans work great. You might not like the idea of wiping your kitchen counters with a rag that had once wiped up a pile of cat puke even if it's been washed. If so, you can always separate your rags into three piles, grab three different colors of permanent markers, and label one pile "bathroom," another pile "kitchen," and the last pile "pets." Or put different colored dots on each categorized pile. This will keep your rag collection organized in a way that ensures you don't use the same rag you once cleaned the toilet with to clean the coffee table.

And if you're thinking, "Sounds great! But my family would NOT be on board," show them the savings! In my household, we save $100 to $200 each year. Simply stop buying paper towels. Everyone in your house will have no other choice but to use cloth once they disappear.

ALTHOUGH SUPER CONVENIENT, PAPER WASTE LIKE DISCARDED PAPER TOWELS ACCOUNTS FOR OVER 25 PERCENT OF TOTAL LANDFILL WASTE.

COFFEE FILTERS

If you make coffee at home every day, you're awesome! You're already avoiding a ton of disposable coffee cups, lids, sleeves, and stirrers as well as saving around $5 a day or more. Home brew, though, usually comes with a disposable filter. Regardless if it's paper, it's still an item that is purchased and then tossed.

A lot of newer coffee machines have mesh inserts, so you don't need a paper filter. If you're in that category, you can go ahead and skip to the next section. Yay, you!

Instead of using paper filters, try making a reusable one from scrap fabric. The zero waste game is all about being resourceful. But if you don't want to sew anything, you can just lay cloth over the area where the filter should go. Or another idea is to use coffee filters made from recycled content that you can compost after use.

I'm not saying that you can't buy anything. In fact, there are several options for reusable coffee filters that you can purchase online. Rule of thumb: support companies that are true to the values of caring for our planet and the workers who make the products.
Also, try to find companies that are close to you in proximity. Reducing the miles any product has to travel is another way to reduce the purchase impact.

PAPER NAPKINS

Instead of disposable napkins, use cloth. Cloth napkins come in handy, even outside of the kitchen. I always keep one in my backpack to wipe my hands or blow my nose when I'm on the go. You can also use them to carry food purchased from a food truck or stand (not the same one you use to blow your nose, of course!).

DISPOSABLE PLATES

Use your real dishes. Plain and simple.

If you're wondering if using paper plates is better than wasting water washing real dishes, consider the manufacturing process of creating disposables. White paper gets its color from being bleached, and chlorine compounds are some of the most hazardous pollutants used in manufacturing that can cause health complications when people are exposed long term according to the Centers for Disease Control and Prevention. Paper products aren't good for the environment or our health.

Even though paper plates are paper, they are not recyclable because of the food residue on them. Plus, bleached white paper plates aren't the best for compost given the chemicals. So if you must use a disposable plate, look for non-bleached plates made from recycled content and by responsible manufacturers. Then compost them when done.

If you're traveling, stock your zero waste kit with some reusable plates. You can find cute secondhand, reusable plastic plates at a local thrift store. Whatever you do, don't use disposable plastic or Styrofoam plates! Avoid these like the plague.

DISPOSABLE CUTLERY

Use real cutlery and wash it right away. If you're like me and you hate doing dishes, wash them as you go to avoid a mound of dirty utensils in your sink.

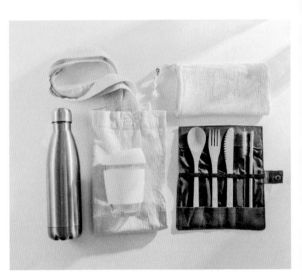

PLASTIC WRAP

That roll of plastic wrap in your kitchen is still plastic, but you can use it wisely. Rip off the exact amount you need, cover your bowl, and discard the wrap after use. No mess, no hassle. True, plastic may be convenient, but there's also a cost associated with this convenience. Here are some alternative, low-cost solutions.

Reusable Food Storage Containers: Put leftovers in food containers with lids. If you run out of food containers, use mason jars or reuse food jars from things like pickles, mayonnaise, and jam.

Creative Covers: Ditch the plastic and instead put a plate on top of a bowl or a bowl on top of a plate. Sometimes, you can even store food without any sort of wrap or container. For example, if you have a watermelon wedge or sliced orange, put the fruit flesh side down on a plate and right into the fridge.

Reusable Food Packaging: There are several single-use containers and bags that food already comes in that you could save for reuse later. You probably have a few plastic containers from restaurant leftovers lying around. Save them for food storage or even for packaging some cookies for a friend. But don't heat up any food in plastic, as the by-products and nasty chemicals that make up the plastic container can leach into the food.

Wax Wraps: There are reusable wax wrap alternatives that work just as well as plastic wrap. Reusable wax wraps are pieces of cloth coated in beeswax or a plant-based wax. You can wrap them over a bowl, plate, or around the food itself for prolonged food storage.

Bowl Covers: You can buy cloth bowl covers or make them yourself using pretty patterns with easy instructions you can find online. Bowl covers that are made from cloth come in various sizes and have elastic around the edges, which allows them to tightly cover bowls, dishes, and plates.

SANDWICH BAGS

Especially if you pack lunches for kids every morning, it's going to be difficult to ditch prepackaged snacks and disposable zipper bags. I will fully dive into how to prepare zero waste lunches later in this book. For now, here's a list of disposable sandwich bag options:

Reusable Containers: If you have small children and you'd rather them not carry a lunchbox full of glass, use smaller plastic containers.

Reusable Sandwich Bags: This eco-friendly option comes in bright colors and fun patterns—and it's a great way to rethink what you purchase and reduce waste.

Paper Sandwich Bags: If the above still doesn't work for you and your family, look for some plastic-free, paper options and responsibly take care of the bag after use.

TRASH BAGS

Zero waste or not, you're still going to produce trash, and trash bags are something we don't typically think about when it comes to waste reduction. Thankfully, there are still ways to nix this disposable and recoup some costs associated with your monthly supplies bill.

The simplest and most cost-effective choice would be to go without a trash bag altogether. If you compost, then the garbage you have should, for the most part, be dry. Once your bin is full, dump it into your curbside bin. Periodically, use some natural cleaning spray to wipe out the bin.

However, I would caution against this method if you still produce quite a bit of garbage only because loose garbage can be messy and find its way out of the trash truck easily. Here are some other items you can use instead of plastic trash bags:

- Large empty pet food bags
- Plastic grocery store bags that can't be recycled anyway
- Washable, reusable trash bin liners
- Trash bags made with recycled plastic
- Large yard waste bags made from recycled paper

TEA BAGS

Who would have thought tea bags have plastic in them? Well, most do. Instead of buying tea bags, try loose-leaf tea. I keep my various teas in pint-size mason jars in my tea cupboard along with honey, my favorite mugs, tea strainers (you'll need one of these for loose-leaf tea), and various spices like cinnamon and turmeric.

You could also make your own reusable tea bags for loose-leaf tea if you don't have a strainer. Again, this is another beginner sewing project that could also double as a wonderful gift. In fact, gifting loose-leaf tea to my friends and family is something I've been doing for years. And if you'd rather not sew, reusable tea bags are available online.

If you do have access to composting, you could research which tea bag brands are 100 percent plastic-free and compost them after use. Another fun way to "zero waste" your tea addiction is to grow your own!

Here's how to make your own beautiful tea garden:

YOU'LL NEED:

- Container big enough for all the plants
- Stones (enough for a layer at the bottom of the container)
- Dirt to fill the container
- Herbs perfect for tea

Your local greenhouse may have different varieties of herbs that could be a good addition to your tea garden.

SOME POSSIBLE CHOICES INCLUDE:

- Lemon Balm
- Peppermint
- Stevia
- Lavender
- Chamomile
- Echinacea
- Rose
- Calendula
- Basil
- Lemon Verbena

HOW TO MAKE TEA WITH YOUR NEW TEA HERB GARDEN

1. Pick a few of each leaf or whatever combination you'd like and steep in a reusable tea bag or tea ball.

2. Or dry your leaves in the sun or in a dehydrator, crush them up, and steep them in a reusable tea bag or tea ball.

COFFEE PODS

I remember when coffee pods made their debut. Of course, I fell prey to the convenient at-home coffee shop machine that could make a variety of fancy drinks. I used that machine so much that I replaced it at least once. But when I dove into my sustainable living endeavor, I quickly realized that coffee pods were contributing to my overall waste.

Coffee is grown in massive fields similar to how corn is grown in the rural parts of the country. Biodiversity like trees and other native plants are removed to increase open, growable square acreage for coffee plants. Of course, this results in loss of important species that help keep insects and other pests away from the plants. Now, pesticides are used to do the same job—a lot of pesticides. So, how can you still enjoy that morning brew while being conscious of the environment, not to mention your health?

REDUCE

A recurring theme to reduce our impact is to reduce consumption. Drink less coffee. Reducing your coffee intake will ultimately reduce your coffee-drinking impact, reduce waste, and save money.

REUSE

Swap your pods for reusable, mesh pod inserts. Add instant coffee or ground coffee to the insert and brew. Many grocery stores even sell loose coffee at in-store grinders where you can bring your own container (think mason jars!) and fill it up with your favorite blend.

RECYCLE

If you don't have access to bulk coffee, look for the better option. Buy coffee in recyclable pods. Many companies are working to make their pods more sustainable, and recyclable aluminum pods are available for most brands.

ROT

If you compost, compostable pods are also available. This is a better option if you just can't kick—or at least limit—your pod drinks.

DISH SPONGES

You know the blue dish sponge with the coarse, scratchy side that is a powerhouse at getting dishes scrubbed clean? Alas, it's plastic and bound to end up in a landfill. There are also the plastic dish wands with fillable compartments, plastic scrubbing brushes, and those crinkly-looking, netted scrubbers that do nothing but capture uneaten food residue. Gross.

Rest assured, there are many sustainable options available that will get the job done. But what's the point of purchasing sustainable options if they're going to end up in the trash? Remember, it's important to take ownership of everything you buy—whether it's a car or a dish brush. A sustainable lifestyle isn't just about purchasing something different. It's about changing your mind-set with regard to consumption.

↓ REDUCE

Resist the urge to keep purchasing plastic sponges and brushes. It's just as easy to take better care of them so they last longer.

↻ REUSE

Washcloths are the most sustainable option, especially if you use one made from scrap fabric. Once the rag is too worn out to use anymore, it can be composted as long as it was made from natural materials.

ROT

There are plant-based dish brushes that work well for scrubbing the hard-to-get food off of dishes. Once these scrub brushes wear out, you can compost them. Don't throw them into the trash! These types of brushes will last a long time if you take care of them. Once you're finished using a plant-based brush, lay it bristle side down so that it can dry and never let it soak in water.

DISH SOAP

Most dish soap ingredients carry harmful chemicals that can stick through cutlery and ultimately be ingested, get absorbed into your skin, and even contaminate your water.

A few common ingredients in dish soap include harmful chemicals like phosphates, sodium lauryl sulfate, diethanolamine, and fragrance. Baking soda works as a cleaning agent for cookery and dishes. It's abrasive enough to get food off, it soaks up oil, and it's okay if you happen to ingest a bit of it. And if you live near a store with bulk options, look for package-free dish soap there. Most bulk options are plant based and more natural.

There are also solid dish soap bars that are available through various online eco shops and natural food stores. Set the block next to the sink, use a cloth or brush to gather suds, and scrub away at those dishes. And if you feel so inclined, you can make your own dish soap. Here's how:

HOMEMADE DISH SOAP

INGREDIENTS:

- 1½ cups (360ml) water
- ¼ cup (25g) grated castile soap (in paper, then compost wrapper)
- 1 tbsp (10g) washing soda (purchase in cardboard, then recycle or compost)
- 1 container for soap (use whatever you have; an old olive oil bottle works great!)

INSTRUCTIONS:

1. Add all ingredients to a pot and heat until everything is dissolved.

2. Wait for the mixture to cool; it shouldn't separate at this point.

3. Transfer the liquid into a bottle with a funnel.

Here are some other ways to reduce waste while doing the dishes:

Scrape Plates Clean: If you compost, this is a great opportunity to make sure every food bit goes to the earth rather than down the garbage disposal.

Rinse and Reuse: Use the same plate, cup, bowl, and utensils all day. Just rinse them clean after each use with hot water. This will keep your kitchen clean when it comes to dishes, and it saves water by not having to fill a sink full of water to wash a days' worth of dishes.

Cook and Eat Out of the Same Pot or Pan: What college student doesn't make ramen on the stove and then eat it out of the pot? Genius, really. If you're having a family dinner, put a large pan or pot in the middle of the table and make it a fun family-style experience.

FOOD WASTE

Humans waste a lot of food on many levels: while farming, during production, at restaurants, in schools, in homes, and more. We could potentially feed every hungry person on earth with the food that goes to waste. And to make matters worse, landfills account for an enormous amount of methane output. Compared to food decomposing naturally, landfills aren't designed to break down food—they only store our waste. They're lined with clay and rubber and covered to prevent liquids from leaching out into the water systems. Since oxygen can't get in and aid in proper decomposition, food breaks down very slowly, allowing bacteria to grow and releasing methane in the process.

Awareness around food waste is on the rise, and even the Environmental Protection Agency (EPA) has developed a food recovery hierarchy to identify and address every level where food waste occurs—saving not only food, but also water, energy, fuel, and labor. Thankfully, there are things you can do to help prevent food waste.

SOURCE REDUCTION

Reducing food production to avoid food waste is the most sustainable option. Many times, farmers will let perfectly good food go to waste as it decomposes back into the ground. It may seem harmless enough when the demand is low, but there are components of the farming process that are affected when food isn't sold.

What can you do to help the situation? Pare down the amount of food you purchase, exercise portion control, and reduce excess overall. Although leftovers are good for lunches and additional meals throughout the week, try to cook only the food you and your family can eat. Additionally, when you serve yourself food, only put what you'll eat on your plate. If you tend to overfill a plate, use smaller plates!

FEED THE HUNGRY

Every year, millions of households around the world can't provide enough food for all their members due to a lack of resources. Yet, millions of pounds of food are thrown away each year.

When there is food left over, get that food to people. Whether it's donating the food to a food pantry or taking leftovers to a family member, help get that food eaten. Better yet, you can have a "get-together" with family and friends. Once the evening comes to an end, divide leftovers with those who attended.

FEED ANIMALS

The next tier in the hierarchy is to feed excess food to animals. And I don't mean sneaking Fluffy a scrap of food

NEARLY ONE-THIRD OF THE FOOD WE PRODUCE IS WASTED—THAT'S ABOUT 133 BILLION POUNDS.

from your plate. But if you have chickens, they LOVE most produce scraps, and this saves money in animal feed, too.

The EPA recommends that businesses donate excess food waste to farms with livestock rather than having it hauled off to the landfill. In most situations, it's cheaper. It is also important to note that states have different regulations regarding donated food, so if you're interested in doing this, be sure to read your state's fine print first.

INDUSTRIAL USE

Excess food can also be used for energy. Using food waste as an alternative fuel reduces our dependence on fossil fuels, while alleviating some of the problems associated with food waste. Many facilities across the country have already taken advantage of this. When the above tiers of hierarchy aren't an option, compost excess food waste to allow the food to naturally decompose back into the earth.

FOOD EXPIRATION

There are also many ways we can prevent food waste in our homes by being aware of illogical food expiration dates. Managing the food—meat, produce, dairy, and bread—in your kitchen by expiration date helps ensure that it gets used and eaten before it goes bad. Any canned or jarred foods, condiments, grains, and dry goods usually last for a very long time without any urgency to use it first. However, here are some expiration labels on food that can be misleading:

Best If Used By: This is not an expiration date. This is merely a recommendation from the manufacturer for when the food tastes the best.

Use By: This, again, is a recommended time to use the product for best quality.

Sell By: This tells the store how long to display the item, but it doesn't indicate that the product is bad after this date.

Even the U.S. Department of Agriculture (USDA) recognizes that food is edible after these dates, but use your common sense. If the food looks, smells, or seems weird in any way, compost it.

COOK WITH SCRAPS

People tend to waste more food more often because it seems to be abundant and easy to come by. But what if you used more of your food before it made its way into the compost bin? Here's how you can maximize your produce.

Carrot Tops: Turn them into pesto! Who says basil is the only leafy green that can be made into pesto?

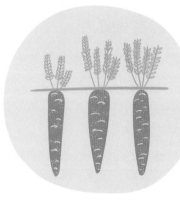

CARROT TOP PESTO

INGREDIENTS:

- About 2 cups (40g) carrot tops
- ½ cup (75g) cashews
- ¼ cup of nutritional yeast
- ½ cup or so (120ml) olive oil
- Salt and pepper to taste

INSTRUCTIONS:

1. Clean the carrot tops.

2. Add the carrot tops, cashews, and nutritional yeast to a food processor.

3. Pulse the food processor until the ingredients are roughly chopped.

4. Slowly add the olive oil while blending until smooth.

5. Add the salt and pepper and blend.

6. Serve with pasta or on top of crusty bread.

Broccoli Stems: Make broccoli chips. It's a win-win when you are reducing food waste and eating healthy food.

Citrus Peels: Use leftover citrus peels from lemons, limes, or oranges and infuse with vinegar to make a zero waste, all-purpose cleaner. See the recipe on page 125.

Banana Peels: Before you toss them, use them to shine you plants or shoes! You can also rub them on bug bites to stop the itch.

Apple Cores: Why not just eat them? Or you could save them to make your own apple cider vinegar.

BROCCOLI CHIPS

INGREDIENTS:

- Broccoli stems
- Olive oil
- Salt and pepper

INSTRUCTIONS:

1. Preheat the oven to 400°F (200°C).

2. Slice broccoli stems thinly into disks.

3. Spread them out on a baking sheet—use a silicone baking mat to reduce parchment paper waste.

4. Sprinkle them with the olive oil, salt, and pepper and toss to coat.

5. Bake for 30 to 40 minutes, or until crispy.

APPLE CIDER VINEGAR

INGREDIENTS:

- Apple peels and cores
- 2 tbsp (25g) of sugar
- 2 cups (480ml) water, preferably filtered

INSTRUCTIONS:

1. Add the apple scraps to a quart mason jar. Fill it about three-fourths of the way.

2. In a measuring cup, dissolve the sugar in the water.

3. Pour the water/sugar mixture into the jar with the apple scraps.

4. Use a fermentation weight or another smaller glass jar to push the apple scraps down below the water. We don't want any apple scraps touching air or they could mold.

5. Cover the top of the jar with a thin piece of cloth and attach with a rubber band.

6. Put the jar in a dark place to rest for 2 to 3 weeks.

7. After that time, strain out the apple scraps (compost them).

8. Return the jar back to the dark spot for another 2 to 3 weeks, stirring every few days.

Potato Peels: Roast them with a little bit of olive oil and salt. They are a yummy, zero waste snack that curbs those chip cravings.

Lemon Peels: In addition to using them to make all-purpose cleaner, you can freeze them in smaller chunks and use them to clean the garbage disposal.

Watermelon Rinds: Pickle them! This might be a Midwestern thing, but they are pretty yummy.

Veggie Scraps: Save onion peels, herb stems, and other veggie bits in the freezer until you have enough to make a vegetable stock.

Stale Bread: You can use stale bread to make everything from croutons and breadcrumbs to delicious bread pudding.

Coffee Grounds: Grounds make great facial scrubs. See the recipe on page 38.

STALE BREAD CROUTONS

INGREDIENTS:

- 2 cups (100g) stale bread cut into 1-inch (2.5cm) cubes
- 2 tbsp (30ml) olive oil – until all cubes are coated
- ½ tsp garlic powder
- ½ tsp paprika
- Salt and pepper to taste

INSTRUCTIONS:

1. Preheat the oven to 375°F (190°C).
2. Add the stale bread, oil, and spices to a bowl and mix to combine.
3. Spread the croutons on a baking sheet evenly.
4. Bake for about 10 minutes, or until your desired crispiness.

WATERMELON RIND PICKLES

INGREDIENTS:

- 2 pounds (910g) watermelon rinds
- ¼ cup (60g) pickling salt
- 4 cups (960ml) water
- 1 cup (240ml) white vinegar or apple cider vinegar
- 2 cups (400g) sugar
- 1 tsp chopped ginger
- 1½ tsp whole cloves
- 2 cinnamon sticks, broken up
- ½ lemon, sliced

INSTRUCTIONS:

1. Cut the rinds into 1-inch (2.5cm) pieces.

2. Add the rind pieces and pickling salt to a large (1 gallon [4L]) container with the water.

3. Let the rinds soak overnight.

4. Rinse and drain the rinds.

5. Add the rinds to a pot and simmer on the stove with just enough cold water to cover the rinds.

6. Cook the rinds for 30 minutes or until tender.

7. In another pot, simmer together the vinegar, sugar, ginger, cloves, cinnamon, and lemon for about 10 minutes.

8. Strain the watermelon rinds.

9. Strain the vinegar/spice mixture, saving the liquid.

10. Add the rinds and vinegar/spice mixture back to the pot to simmer for another 30 minutes, just until the rinds are translucent.

11. Add the rinds to jars, let them chill for a few minutes, then eat or can them for later.

COFFEE GROUND FACIAL SCRUB

INGREDIENTS:

- ¼ cup (25g) coffee grounds
- ¼ cup (50g) brown sugar
- ¼ cup (60g) coconut oil

INSTRUCTIONS:

1. In a bowl, mix the coffee grounds, brown sugar, and coconut oil together until well combined.

2. Use to scrub your face.

3. Rinse off with cool water.

COMPOSTING

Composting can seem intimidating at first, but if your primary goal is to keep food waste out of the landfill, it's a lot easier than you may think. When left to do its thing, organic waste will eventually break down back into the earth.

If you have your own yard space, the options are endless. My favorite way to compost is in an open pile enclosed with either wood or wire. You can build a compost enclosure with nearly any material you have available to you, like scrap wood, old pallets, or even chicken wire wrapped around posts. Once you have an enclosed area, you dump your scraps into the pile and let nature do its thing. The hardest part is walking from the house to the compost outside.

As far as compost enclosure placement goes, I recommend a few things:

- Place the enclosure in an area in your yard away from your home. Compost can get very hot and could potentially cause a fire if not maintained properly.

- Look for a place that gets equal sunlight and shade. If piles get too hot or too wet, they will not break down properly.

- Make sure your pile is within reach of a hose because you will need to moisten it occasionally to keep the process going.

If you don't have shade, simply cover your compost so that it doesn't get too dry. And if you have a lot of shade, you will not have to moisten it very often.

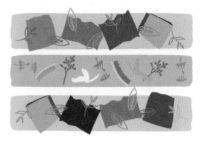

If you really want to get your compost decomposing rapidly and get that perfect, rich, black soil, add compost before you add anything else to your pile. Get some from a friend, a community garden, or buy a bag of compost from your local garden store. You can compost without this step, it just speeds the process and gets those necessary microbes established and ready for whatever you start dumping in later.

When you start adding compost, layer your "greens" to "browns." This means add an equal part "browns," which is materials that come from trees like paper, cardboard, leaves, and sawdust, to "greens," which is everything else.

Here's my "everything else" list of what I tend to add to my pile:

- Fruits and Veggies
- Eggshells
- Nuts and Seeds
- Cardboard and Paper (browns)
- Coffee Grounds
- Tea Leaves and Tea Bags (make sure they are plastic-free since a lot of bags contain plastic)

- Yard Waste
  - » Clippings
  - » Leaves
  - » Branches
  - » Wood Chips
  - » Sawdust
  - » Weeds
  - » Potted Plants
- 100% Natural Fibers
  - » Cotton
  - » Wool
  - » Coconut
  - » Manila
  - » Sisal
  - » Pina
  - » Flax
  - » Hemp
  - » Jute
  - » Ramie
  - » Cashmere
  - » Wool
  - » Alpaca
  - » Mohair
  - » Silk
- Houseplants
- Hair and Fur
  - » Dog
  - » Cat
  - » Human
- Dust Sweepings and Vacuum Dirt

I recommend burying things like meat, bones, dairy, and oils. And if you still aren't sure if you can compost something, an internet search will usually answer your question.

Once you start layering your compost, use a shovel to turn the compost mixture periodically and make sure it has the perfect balance of moisture. Nature will do the rest. If you're worried about critters getting into your pile, cover it with a lid. You can also reinforce the bottoms with some wire to make sure possums, raccoons, rats, or stray animals can't find their way into your food scraps.

And if you're worried about the smell, compost that is being maintained properly will not have a foul odor. I promise you it will smell like rich, earthy soil. The only time compost will start to reek is when it gets too wet and starts to mold. This isn't good, as it could potentially kill off all those beneficial microbes that are breaking down the organic waste. If you think your compost is getting too wet, add some more "brown," dry components to your pile to soak up the added moisture.

Also make sure you give your refuse as much surface area as possible before adding it to your bin. Shred that paper, cut down those cardboard pieces, break sticks into tiny pieces. This will help those items break down quicker because bigger pieces take much longer to decompose.

There are other options for people who don't have a yard or live in a rental or an apartment but still want to compost.

Compost Bins: If you have a patio or balcony area, you probably have enough room for a small compost bin. Look for a used bin at local secondhand shops or online secondhand marketplaces. And if you have a plastic storage tub lying around, drill some holes in the lid, layer in some "browns" and dirt and start adding your scraps. Soon you'll have rich compost that you can either use in houseplants, give to friends or neighbors, or dump in a green space.

Vermicomposting: This type of composting uses worms, usually red wigglers, to break down organic waste into rich, black soil. Keeping a worm bin is pretty low maintenance as long as it's kept at a steady temperature that isn't too hot or too cold and the worms are not inundated with too much to break down. The materials left over after worms have broken down organic waste is like pure gold, providing gardens and plants with a super nutrient-rich pick-me-up.

Bokashi: This type of composting uses an anaerobic process that breaks down food waste with inoculated bran that ferments the waste—even those

"no compost" items like dairy, meat, and bones. In a closed bucket, the bran and food waste are layered until the bucket is full. It's then sealed to break down for about ten days. By this time, the food should be broken down into "pre-compost," which means it's not fully broken down but in a very fermented, acidic state that is ready to be buried in a free spot in a garden. When the bokashi compost is breaking down, it will need to be drained of the liquid or leachate that is created as a by-product of the fermentation.

Compost Sharing: If you don't have time to do the composting yourself, there are many online spaces that bring people together in your community who compost or who

are looking for a place to compost. Or if you know someone who does compost, consider asking him or her if you can add your food scraps to their pile.

Local Community Garden: Many communities have public or private spaces where gardens are maintained by a group of people to grow vegetables, fruit, or even flowers. These community gardens usually have composting areas, and chances are they'd be happy to take additional compost to nourish the gardens.

Compost Drop-Off: Does your community have a compost drop-off location? If so, keep your compost in the freezer until you've accumulated a big stash to drop off. But first, check with your drop-off location about which items they may or may not accept.

Pickup Service: A lot of communities have compost pickup programs that will sell you a bucket to fill with food waste and then pick it up when it's full. This service usually comes with a monthly subscription and may offer free bags of compost.

Local Farms: The farms in your area sometimes love getting extra compost for livestock or gardens. Ask around and see if you can start delivering edible goodness to these farms to divert food waste from the landfill. Remember, if composting isn't an option, reduce food waste as much as possible before it gets to this step.

FOOD STORAGE

Properly storing food could increase the shelf life and prevent unnecessary food waste. There are several ways to store your food.

Freeze leftovers to prevent them from going bad. Food can be frozen in glass just as well as in plastic. But if you are freezing liquids in glass, be sure to give it plenty of headspace for the food to expand while it freezes (about 2 to 3 inches [5 to 7.5cm]). Keep the lid loose and don't pack the jars too close to each other. You can also pack leftovers in lunch-size portions, so they are ready to "grab and go" for work or school the next day.

Store dry goods in clear containers. When I buy dry goods like flour, sugar, rice, beans, snacks, and nuts in bulk, I immediately store them in jars. This seals in the freshness, prevents any unwanted critters getting into my food, and makes it easy to visually see how much of each item I have for when it comes time to go grocery shopping again.

Store and freeze bread. This is probably zero waste sacrilege, but I keep my bread in a plastic, sealable bread box I got at the thrift store. It works the best, what can I say? I freeze any extra bread I don't plan to eat for a while.

Use or store produce right away.

Since produce goes bad faster than anything else, a little extra care is needed to increase freshness and quality. I try to prepare produce as soon as I get it home from the store. But since I can't always do this, I chop and freeze things like peppers and onions. Here's how to keep other types of produce:

- **Bananas**

  Don't store bananas in any sort of bag once you get them home. Let them sit out at room temperature. If they start to over-ripen before you can eat them all, freeze them for later. You can use them in breads, muffins, smoothies, and dairy-free ice creams.

- **Apples**

  Keep apples in the crisper drawer in your fridge to extend their life.

- **Tomatoes**

  Keep tomatoes at room temperature in the pantry away from sunlight.

- **Carrots**

  Chop carrots, pack them in a jar or any container for that matter, cover with water, and store them in the fridge to keep them crisp for a long time.

- **Celery**

  Like carrots, chop celery and store it in the refrigerator submerged in water.

- **Berries**

  Wash them in a vinegar and water bath to kill any spores that could cause the berries to mold quicker. Then dry the berries well and store them in a jar in the refrigerator.

- **Leafy Greens**

  Wash and thoroughly dry the greens. Then wrap them in a towel and store in the crisper drawer of the fridge.

- **Potatoes**

  Store them at room temperature in a dark place. For example, I keep mine in a basket in the pantry.

- **Onions**

  I store them in their own basket at room temperature in the pantry as well. If you have too many to store, chop and freeze them for later.

- **Peppers**

  You can chop peppers, store them in jars, and keep them in either the freezer or fridge.

- **Avocados**

  Keep avocados in the fridge to slow ripening. Then when you're ready to eat them, place avocados on the counter to speed things up.

CHAPTER 3

GROCERY SHOPPING,

MEAL PLANNING,

AND

RECIPE IDEAS

The waste you generate all stems from what you bring into your household every day. It depends on your access, location, financial situation, and diet. By no means am I saying you need to live a "perfect" lifestyle, but I'm showing you another way to get closer to zero.

-<<<<<<<<<<<

When I did my trash audit, I found that 95 percent of my waste was from things related to food: fast food containers, boxes, plastic film, plastic tubs, chip bags, and other packaging from processed junk I didn't need. I used to eat a lot of meat, cheese, potatoes, and processed foods. Sadly, it showed, and it wasn't until I started getting really concerned about my health that I connected the dots. Think of it this way: Everything that is bad for the planet is potentially bad for your health.

I knew the first step was to change my diet. I started eating fresh, whole foods that either aren't packaged or have minimal packaging, like fruits, veggies, items in the bakery, and foods from the butcher counter.

The cleanest foods with the fewest processed ingredients are housed in the perimeter of the grocery store and in bulk bins. Bulk is food that doesn't have packaging and can be scooped out of containers. Bulk can be anything from dry goods to oils and vinegars to soap. The bulk bins are generally in the health food section of a grocery store. There are even bulk co-op shops that focus on natural, healthy, and local food.

GROCERY SHOPPING

To make grocery shopping less overwhelming, below is my go-to list of what I bring to the store and how to keep it all organized. You'll also find more alternative ways to reduce shopping and food packaging waste if you don't have access to bulk, package-free food after my short go-to list.

① Cloth Grocery Bags

I take three or four cloth grocery bags with me to the store, which is always enough for my weekly grocery haul. You might need more or less, depending on how big your family is and how often you shop.

② Cloth Produce/Bulk Bags

Cloth produce bags are essential items for avoiding single-use waste. I use them for produce, items out of the bulk bins, and baked goods! Online retailers usually have several options available for purchase. You can also make some from scrap fabric or old clothing, sheets, or curtains.

Some people bring a charcoal pencil with them to write price look-ups on their bags, but I just use the memo app on my phone and give them to the cashier at checkout.

❸ Jars

There are a few wet items I get while I shop, including peanut butter, almond butter, dish soap, laundry soap, and oils. Jars make for an easy, airtight container to transport your goods home. I also use jars for spices and baking supplies like baking soda and cocoa powder. Keep in mind that you do NOT pay for the weight of the jar. The cashier will subtract the tare weight, which is the weight of the empty jar, from the total weight of the jar and items at checkout.

If you eat animal products, you can also use jars to get meat and cheese from the butcher and deli counters. These work much better than the paper they usually wrap the products in, considering jars won't leak!

ONCE YOU'RE MORE TUNED IN TO WHAT YOU'RE LOOKING FOR. YOUR PERSPECTIVE ON FOOD WILL BE DIFFERENT.

❹ Cloth Wine Bag

I find that this little bag is one of my favorite things to keep me organized, which is the key to zero waste grocery shopping. I use it to organize my other shopping components like jars, smaller bags, and wine, of course. Check out your local liquor stores or liquor departments in your grocery stores for these bags.

I keep these cloth bags and jars in my pantry, and when it's time to go to the grocery store or farmer's market, I just grab them and go. When I return from my shopping trips, I make sure to wash the bags clean, including the big cloth grocery bags.

Where to shop can sometimes be a hurdle, but I challenge you to go to your local grocery store without any intention to buy and just browse. Keep your eyes peeled for package-free foods. You'll be surprised at what you can find—everything from bins of loose trail mix and granola to beans and rice. And don't worry if you don't have access to bulk. Most people don't. Again, it's not about living a ZERO waste lifestyle, it's about reducing.

AVOID PLASTIC BAGS

You don't need them. You should be taking your produce home to wash anyway, so there's no need for them other than to keep a larger quantity of items together like a pound of potatoes or a bag of peas. In this case, bring your own reusable cloth produce bags that are more durable anyway.

AVOID PLASTIC PACKAGING

If you can, avoid products that are primarily packaged in plastic. For example, opt for the box of cornstarch rather than the plastic container.

BUY THE LARGEST CONTAINER

Yes, you can sometimes save money buying products out of the package-free bulk bins, but you often end up spending more money. Buying in larger quantities also eliminates several smaller containers

of products ending up in the landfill. Keep in mind that I'm referring to items that have longer shelf lives.

EAT A MORE PLANT–BASED DIET

Cutting more animal products out of your diet has many health and environmental benefits. Most animal products are packaged in nonrecyclable, plastic packaging that's usually tossed after you prepare dinner. By making fruits and vegetables a bigger part of your diet, you not only decrease your trash waste, but you also contribute to overall energy, water, and greenhouse gas reduction.

GROW AND CAN YOUR OWN FOOD

If you have space, start a garden! Gardening will save you money, eliminate a lot of packaging waste, and is pretty darn rewarding. Canning your excess garden produce is another fantastic way to stock up for the year and reduce food packaging waste.

A DIET OF PLANT–BASED FOODS CAN REDUCE TRASH. WATER WASTE. ENERGY CONSUMPTION AND GREENHOUSE GAS EMISSIONS.

Don't have the outdoor space? Then keep pots of herbs on your patio or deck to access all summer long. Many herbs can be brought indoors during the winter and can be dried.

MAKE MORE FOOD FROM SCRATCH

While there are plenty of prepackaged foods you can make with little waste, there are things like salad dressings, ketchup, and mustard that are easy enough to whip up. You can also get your spouse and kids involved on taco night by having them help make homemade tortillas (see the recipe on page 63). Getting back in the kitchen and cooking with your loved ones has more than one benefit.

ASSESS PACKAGE—FREE OPTIONS

Identify the package-free items in your store like fruits and vegetables. Here's a tip: Take pictures of your bulk sections so that you can reference the photos later for when you are meal planning. And for items that aren't available without plastic packaging, you can opt for metal, paper/cardboard, or glass containers.

PARE DOWN YOUR MEAL REPERTOIRE

I'm sure you've read blogs and articles about how your life is less complicated with less stuff. Well, the same applies to meal planning! It's more manageable to have a handful of meals every week to choose from, and this makes grocery shopping a lot easier, too.

Plus, simple meals require less ingredients, less money, and less stress throughout the workweek. Consider rotating the same meals to reduce food waste. This way, you'll use the same ingredients for meal prep and not have to reinvent the wheel by spending more on recipes you haven't yet tried.

STICK TO A THEME

Pick a few "theme nights" during the week, but change the recipe slightly each time for variety. For example, our family does "Stir-Fridays," and I don't always use the same recipe. I buy veggies that are on sale or in season and incorporate what I already have into the dish.

MEAL PREP

Coming up with your meals ahead of time as well as making a list of ingredients you need is crucial to avoiding waste. Also, once you get your ingredients, prepare what you can ahead of time (i.e., chopping, making rice, baking). Planning meals for the week saves you time, money, and waste—and having meals at the ready cuts down on the temptation to get a fast food meal. I find that when I choose elaborate meals that call for proprietary ingredients, I end up with a pantry full of items that I never use again.

That's another great thing about shopping in the bulk section; if you do want to try something new, you won't end up with a giant package. You'll only get the amount you need.

It's also important to keep some zero waste pantry staples to use throughout the week. Keeping an essential list of pantry goods was the secret to keeping my grocery shopping easy and cheap, my meals simple, and food waste at a minimum.

TRY SUBSTITUTIONS

Don't be afraid to substitute items for ones you can't get package-free. For instance, if you like eating prepackaged cookies as a snack, try bulk trail mix instead (some of them even have cookie dough balls). Or if you're trying to kick dairy, like sour cream for instance, substitute avocado or salsa.

KEEP AN ORGANIZED PANTRY

I bring my items home from the store in my reusable cloth produce bags. I then transfer them into glass storage jars to keep them fresh. This also lets me see what I have and what I need to replace or fill up. Another key component to a zero waste pantry is stocking up on items that have longer shelf lives. This way, you can buy a larger amount, reducing your need to travel to the store as often. Here are the staple items I shop for and stock up on for meal prep.

Grains

Grains are the base for many meals in my household. My family loves stir-fried veggies, curries, and a hearty bowl of pasta. Quinoa is also great for making veggie "meats."

- Rice
- Quinoa
- Oats
- Pasta

Baking

Baking is such an important part of my life that I gave it its own category on my shopping list. Everything from cupcakes, pancakes, bread, pies, and more comes out of my kitchen quite regularly. And of course, I use baking soda in many of my natural cleaning recipes.

- Flour
- Sugar
- Baking Powder
- Baking Soda

Legumes

Beans are so versatile. They can be used in just about any dish, even dessert! I use lentils as a meat substitute in a lot of dishes like chili and pasta sauce.

- Chickpeas
- Black Beans
- Lentils
- Pinto Beans

Natural Sweeteners

Maple syrup is good for more than just pancakes. It can be used as a honey substitute and replace a lot of refined sugar in dishes. Dates are great for binding ingredients together like in pie crusts or energy bites.

- Maple Syrup
- Dates

Nuts and Seeds

Nuts are so versatile! Almonds are used to make almond milk, peanuts are turned into peanut butter (see the recipes on pages 65 and 67), and cashews are used in a lot of vegan cheese substitutes.

- Almonds
- Peanuts
- Cashews
- Popcorn

Oils and Vinegar

Olive oil is my go-to cooking oil. I use it when I am sautéing veggies or need to lubricate a pan in place of cooking spray. Coconut oil is used in everything from baking to toiletries, and vinegar is the primary ingredient in my cleaning spray.

- Olive Oil
- Coconut Oil
- White Vinegar

Spices

Here's a tip: Simply stock up on your favorite spices. I do have to point out that nutritional yeast is a great Parmesan cheese substitute if you're looking to cut dairy from your diet.

- Salt
- Pepper
- Vanilla Extract or Beans
- Nutritional Yeast
- Garlic Powder
- Parsley
- Cumin
- Chili Powder
- Onion Powder
- Yellow Curry
- Ginger
- Cinnamon

- Paprika
- Thyme
- Oregano
- Basil
- Cayenne

Fruits and Veggies

Most grocery stores offer unpackaged fruits and veggies. You can also look for unpackaged produce at local farmer's markets. A farmer's market is your best option, considering most produce will be local and in season. You could also look into community-supportive agriculture programs or community gardens for package-free goodness. I challenge you to seek out farmer's markets for produce and anything else you can incorporate into your weekly meal plan before heading to the local grocery store.

Dry Goods/Bulk Bins

Take your own containers to the store (you can use cloth produce bags) and fill up with what you need. I get flour, sugar, beans, rice, oats, granola, trail mix, lentils, and more from the bulk bins.

You can sample small amounts of new food without committing to a big bag, which ultimately helps divert millions of pounds of waste from landfills each year. And sometimes you'll spend less on the bulk item versus its packaged twin.

Below are the steps that I take while stocking up on delicious package-free goods.

Locate stores that sell bulk package-free items. Some of us are not as lucky as those who have access to stores with bulk sections, and that's okay! The more we demand to shop bulk items, the more stores will adapt. It's amazing how much power we have as consumers, and finding bulk stores is easier than you would think. Simply, search "bulk food" online and it will generate results.

Bring your own reusable containers or bags to the store. Mason jars work perfectly for bulk nut butter and liquids. For dry goods, you can make cinch bags out of scrap fabric. You can also make bags from stained or torn clothing and old bed sheets and curtains. If you prefer not to make them, a quick internet search will give you plenty of online stores that sell reusable bags.

Get the tare weight of your containers. To avoid paying for the weight of your bags or containers, you will first need to get the tare weight, which is the weight they are when empty. Once you have that amount, write it down and ask the cashier to deduct that amount from the total weight once the container is filled (don't forget to weigh your containers

with the lids on when you get the tare weight). Sometimes you can tare your containers yourself on the scales near the bulk sections or ask customer service to do it for you.

Transfer bulk into your jars at home. At the store, fill your containers with whatever package-free, bulk item you need. Once home, transfer your bulk bounty into jars for storage. Having bulk in jars makes it easy to see what item is getting low and needs to be replenished.

Spices

Most bulk stores and even farmer's markets offer bulk spices. You can use small mason jars or reuse existing spice containers to carry and store them. Either tare your jar beforehand or have customer service at the grocery store weigh your empty containers before you fill them so the jar weight

can be subtracted at checkout—this way, you're only paying for the product, not the packaging.

Bread

Look for bread at your local farmer's markets or a local bakery. You can even ask to buy bread without the wrapper at your grocery store. I sometimes go to local bakery-style restaurants to get fresh bread as well as bagels, rolls, and biscuits. I've recently purchased a secondhand bread machine for three dollars, and baking homemade, package-free bread has never been easier!

Meat, Eggs, and Cheese

If you consume meat and dairy, look for local butchers who farm sustainably. Ask the butcher to fill your metal or glass container. For cheese, use your own containers at a cheese shop or deli counter. Look for milk in glass containers. For eggs, check out your local farmer's market where you can return your empty cartons or befriend a neighbor who raises chickens. I will never judge or berate anyone for what he or she eats. I do suggest that you consume less meat and dairy in general, seeing as this industry is incredibly wasteful. Try implementing more plant-based meals into your routine. You'll experience new flavors and recipes while saving money in the process!

Liquids/Nut Butters

Many bulk stores also have oils, vinegars, syrups, and nut butters in bulk. Again, bring your own containers and fill them up with the amount you need.

Soap/Laundry Detergent

Access is limited to where you can buy liquid soaps in bulk. However, package-free bar soap is more prevalent. Don't stress yourself out if you can't find everything sans the packaging. Look for items in paper or cardboard and recycle.

Have fun exploring your local stores to see what you can get zero waste, but stay strategic and only buy the things that you truly need. Just because you can get something zero waste doesn't mean you should. Remember to keep your consumption in check.

And finally, think about storage. Take a look around your home for glass containers. Mason jars and glass food jars make perfect bulk food storage. You don't want to leave it in the cloth bags you scooped it into because the food might go stale or attract critters. I recommend repurposing before purchasing jars. If you tend to get pasta sauce or pickles, use those jars after they are empty and cleaned. If you're in need of more, check thrift stores for mason jars or see if family or friends have any to spare.

MEAL PLANNING IDEAS

Coming up with your meals ahead of time as well as making a list of ingredients you need is crucial to avoiding waste.

Here are some meal plans to get you started on creating a zero waste plan. Feel free to make substitutions, add, or even do without certain ingredients. The goal is to make each meal as low waste as possible.

Each meal includes plants, and I've tried to be inclusive on what I think most people would have access to without too much packaging.

Week 1

MONDAY

BREAKFAST

Oatmeal with Seasonal Fruit

Shop the bulk bins for the oatmeal, use homemade nut milk, and add seasonal fruit to make this breakfast near zero waste and packed full of goodness.

LUNCH

Kale Caesar Salad with Crunchy Chickpeas

Toss a kale salad with a creamy, homemade Caesar dressing made with lemon, Dijon, garlic, and a little tahini. Top the salad with roasted chickpeas seasoned with garlic, salt, and pepper. Add a piece of homemade garlic bread.

DINNER

Spaghetti with Lentil Bolognese and Homemade Garlic Bread

Cook bulk pasta (or homemade, you rockstar!) and top with store-bought pasta sauce (in glass) combined with lentils bought in bulk. Pair this with leftover kale salad from lunch.

TUESDAY

BREAKFAST

Smoothie

Use seasonal and package-free produce with ice and homemade plant milk (see the recipe on page 65) to make a yummy breakfast smoothie.

LUNCH

Leftover Spaghetti from the Dinner Before

Make sure there is NO food waste for this meal plan to be zero waste. When you have leftovers, always pack them for lunch the next day.

DINNER

Black Bean and Sweet Potato Tacos

Roast a sweet potato with chili powder, cumin, paprika, and garlic. In a skillet, roast peppers, onion, and garlic until tender with the same spices you used for the sweet potato. Add black beans to the skillet. Serve with homemade tortillas and top with cilantro.

WEDNESDAY

BREAKFAST

Kicked-Up Toast

Kick your toast up a notch with homemade or package-free peanut butter (see the recipe on page 67), fresh fruit, or smashed avocado with a little garlic.

LUNCH

Fiesta Bowls

Use some of the leftover taco filling from Tuesday's dinner and pair it with cooked quinoa. Top the dish with salsa and cilantro.

DINNER

Veggie Stir-Fry

Sauté peppers, broccoli, and peas with rice wine vinegar, soy sauce, juice from a squeezed orange, and some grated ginger. Serve on top of a bed of rice.

THURSDAY

BREAKFAST

Homemade Banana Muffins

(See the recipe on page 64).

LUNCH

Leftover Stir-Fry

Heat up leftover veggie stir-fry from Wednesday night's dinner.

DINNER

Chili with Beans and Lentils

In a pot, sauté peppers, onions, and garlic. Add beans, lentils, and crushed tomatoes with their juice. Add a little veggie stock to thin it out. Top with cilantro.

FRIDAY

BREAKFAST

Omelet with Spinach and Salsa

For the salsa, add roma tomatoes, onion, green bell pepper, a jalapeño, garlic, and spices to a food processor. Pulse until your desired consistency.

LUNCH

Leftover Chili

Heat up chili from Thursday night's dinner and put over a bed of rice.

DINNER

Homemade Pizzas

Make this a family occasion! You can make your own pizza dough, make or buy sauce, and top with all of your favorite package-free veggies. If you want to add meat, have the butcher put it in your own containers.

SATURDAY

BREAKFAST

Pancakes

See the recipe on page 66. Serve with fruit and maple syrup. Peanut butter goes on them quite nicely, too.

LUNCH

Chickpea Salad Sandwiches

Mix chopped chickpeas, red onion, celery, mayo, mustard, salt, and pepper. Place between slices of bread.

DINNER

Night Out!

Treat yourself to a dinner out using all the zero waste tips you've learned thus far.

SUNDAY

BREAKFAST

French Toast

Lightly soak slices of bread in an egg, nut milk, vanilla, and cinnamon mixture. Fry in an oiled skillet until crispy. Serve with fruit and a drizzle of syrup.

LUNCH

Leftover Chickpea Salad Sandwiches

Put leftover chickpea salad from Saturday's lunch between two fresh slices of whole-grain bread.

DINNER

Nachos with Black Beans

Top nachos and black beans with lettuce, salsa, chopped green pepper, radishes, avocado, and cilantro. I get my package-free tortilla chips from my local Mexican restaurant.

MONDAY

BREAKFAST

Overnight Oats

In a pint jar, add ½ cup (40g) oats, ½ cup (120ml) nut milk, and 1 tablespoon (8g) chia seeds. Put the lid on and let it sit in the fridge overnight. In the morning, top the oats with fruit and nuts. Add maple syrup or a dollop of peanut butter.

LUNCH

Salad in a Jar

I love meals in jars, especially for work. They fit nicely in my bag and make cleanup so easy.

DINNER

Kale and White Bean Soup

Sauté onions, carrots, and garlic in olive oil until fragrant. Add salt, pepper, thyme, and sage. Add veggie stock, white beans, pasta, and kale. Cook for about 20 minutes or until the carrots are tender and the noodles are cooked through. For the pasta, I use whatever I have, even if it's frozen spaghetti that I store when I make too much. But also feel free to make homemade noodles or even dumplings.

TUESDAY

BREAKFAST

Banana Smoothie

Use those overripe bananas in your freezer mixed with ice and some yogurt.

LUNCH

Salad with Kale Bean Soup

Heat up kale bean soup from Monday night's dinner and pair with a fresh salad.

DINNER

Roasted Veggie Bowls

Roast veggies with olive oil, garlic, salt, and pepper. Make quinoa. Top the quinoa with the roasted veggies, your favorite beans, and drizzle with a sauce made by putting 2 cups (40g) cilantro, 4 cloves of garlic, jalapeño, and olive oil into a food processor and blend until smooth.

WEDNESDAY

BREAKFAST

Muffin and Smoothie

Baking some homemade banana or blueberry muffins makes a quick breakfast option for the rest of the week. Pair a muffin with an easy breakfast smoothie.

LUNCH

Leftovers or Salad

For my go-to salad, I chop some romaine, red onion, and green pepper. Then I add black beans and shredded carrot. I top it with a homemade vinaigrette of oil, red wine vinegar, Dijon mustard, a little honey, and salt and pepper.

DINNER

Eggplant Parmesan

Use whatever pasta you have, jarred sauce, and some sliced eggplant, breaded and fried. Top it with homemade, vegan Parmesan for a comfort meal that will push you over that midweek hump.

THURSDAY

BREAKFAST

Avocado Toast

Top your avocado toast with a slice of tomato and red onion.

LUNCH

Leftover Eggplant Parmesan

Heat up some leftover eggplant Parmesan from Wednesday night's dinner.

DINNER

Veggie Kabobs

Skewer peppers, onions, potatoes, and mushrooms and grill until tender. Use any leftover cilantro/jalapeño sauce from Tuesday's dinner to top.

FRIDAY

BREAKFAST

Oatmeal

Top your oatmeal with brown sugar or your favorite nuts and dried fruit. Pair with a muffin.

LUNCH

Leftover Grilled Veggies

Pair leftover grilled veggies from Thursday night's dinner with any quinoa or salad from previous nights.

DINNER

Potato Soup with Toasted Bread and Kale Pesto

To make the soup zero waste, find ingredients in glass or tin rather than plastic, and be sure to recycle after. Get some crusty bread from a local bakery and use some of the leftover kale from earlier in the week to make pesto.

SATURDAY

BREAKFAST

Muffin with Veggie Scrambled Eggs

Choose to get your eggs local from farmers who take extra care of their chickens.

LUNCH

Leftover Lunch

Eat whatever leftovers you find in the fridge to get it as empty as possible before filling it back up.

DINNER

Pasta with Vegan or Non-Vegan Alfredo

Add tomatoes and roasted Brussels sprouts to your alfredo sauce. If you can't find package-free Brussels sprouts, squash and broccoli are also great options.

SUNDAY

BREAKFAST

Pancakes Topped with Fruit

I always make pancakes on the weekend. It's a tradition. Plus, my son wouldn't be very happy if I skipped.

LUNCH

Family Lunch

It's Sunday! Go visit family and enjoy a meal together.

DINNER

Baked Potato Bar

Some easy, zero to low-waste toppings include: green onions, salsa, tomatoes, red onion, sautéed mushrooms, and roasted veggies. I top my potato with a homemade vegan cheese sauce or a creamy salad dressing from a glass container.

Making a few things from scratch is a great way to avoid packaging, save money, and eat more healthfully. I try to make easy things and prep as much as I can in advance so that a weekly meal isn't taking two hours to get to the table. Do as much meal prep as you possibly can on the weekends to save yourself the hassle during the work week.

And who says that you must be the only one who cooks? Get the whole family involved. If you want to make homemade tortillas but don't have time, assign this task to a child or your spouse. You can even prep the dough balls in advance and keep them in the fridge until taco night.

HOMEMADE TORTILLAS

INGREDIENTS:

- 3 cups (300g) all-purpose flour
- 1 tsp baking powder
- 1 tsp salt
- 1 cup (240ml) water
- ⅓ cup (80ml) olive oil or oil of preference

INSTRUCTIONS:

1. Add the flour, baking powder, and salt to a bowl and combine.

2. Add the water and oil and stir until a dough ball is formed. Use a stand mixer with a dough hook if you have one.

3. On a lightly floured surface, pinch off half-dollar-size balls of dough and roll thin. Use a wine bottle if you don't have a rolling pin.

4. In a medium-hot skillet, cook the tortillas until air bubbles form and the underneath side is browning, then flip. Cook for about 1 minute on each side.

5. Keep cooked tortillas wrapped in a kitchen towel. This is the trick to keeping them pliable.

6. Tortilla dough can be kept in the fridge, in a sealed bowl, for 2 to 3 days. You can also freeze the dough.

INSTRUCTIONS:

1. Preheat the oven to 375°F (190°C) and line a muffin pan with reusable cupcake liners.

2. In a bowl, mix together the flour, cinnamon, baking powder, baking soda, and salt.

3. In another, bigger bowl, combine the mashed bananas, sugar, applesauce, vanilla, and vegan butter.

4. Evenly fill 12 muffin cups.

5. Bake for about 10 to 12 minutes or until you can stick a knife in one and it come out clean.

HOMEMADE BANANA MUFFINS

INGREDIENTS:

- 1 ½ cups (180g) all-purpose flour
- ½ tsp ground cinnamon
- 1 tsp baking powder
- 1 tsp baking soda
- ¼ tsp salt
- 3 medium bananas, mashed
- ⅓ cup (65g) sugar
- ¼ cup (50g) applesauce
- 1 tsp vanilla extract
- ½ cup (120ml) melted vegan butter

There are many homemade muffin recipes you can make with low- to no-waste ingredients.

HOMEMADE ALMOND MILK

INGREDIENTS:

- 1 cup (150g) almonds
- 5 cups (1200ml) water
- 1 tsp vanilla extract
- ⅛ tsp salt

INSTRUCTIONS:

1. Soak the almonds overnight in 2 cups (480ml) of the water.

2. In a high-powered blender, add the soaked almonds, the remaining (720ml) water, vanilla, and salt.

3. Blend until smooth.

4. Strain the mixture through cheesecloth or a kitchen towel (use whatever you have).

5. Keep the milk in a container in the refrigerator for about a week.

6. Don't forget to save the pulp for something yummy later!

HOMEMADE PANCAKES

INGREDIENTS:

- 1 cup (120g) all-purpose flour
- 2 tbsp (25g) sugar
- 2 tsp baking powder
- Pinch of salt
- 1 cup (240ml) almond milk
- 2 tbsp (30ml) olive oil

INSTRUCTIONS:

1. Mix the flour, sugar, baking powder, and salt in a bowl, then add milk and olive oil and stir until completely combined.

2. Warm up a well-seasoned cast-iron skillet or griddle and pour the pancake batter onto the pan to about 5 to 6 inches (12.5 to 15cm) in diameter.

3. When the batter starts to bubble, flip. The pancakes should be golden brown.

4. Once completely done, serve with fresh fruit or whatever toppings you desire and, of course, syrup!

5. Make some in advance and pop them into the freezer for a quick breakfast option during a hectic workweek.

HOMEMADE PEANUT BUTTER

INGREDIENTS:

- 2 cups (300g) peanuts
- 3 tbsp (45ml) olive oil
- 2 tbsp (30ml) agave nectar or honey (optional)
- Salt to taste (optional)

INSTRUCTIONS:

1. Add the peanuts and olive oil to a food processor and pulse until smooth and creamy or to your desired consistency.

2. Add the agave and salt, if using.

CHAPTER 4

BATHROOM,

TOILETRIES,

AND

Beauty

When I started assessing my waste in the bathroom, I noticed that most of the waste consisted of paper products and plastic bottles from toiletries.

‹‹‹‹‹‹‹‹‹‹

You use a lot of disposable products in the bathroom, and it's not hard to amass quite the collection, considering marketing leads you to believe you need a product for literally everything. Before you go and buy a bunch of unnecessary items to replace everything, first look to see if there's something you already own that will do.

Use up what you have, just like with your kitchen items. The goal is to throw less away, so you're only replacing items as they run out. If you do find that there are products that you've already purchased that are unsafe, recycle them if possible or dispose of them. Health always comes before trash. Secondly, reduce what you need. For instance, rather than having a type of lotion for every body part, find one good lotion that will work for most applications. Simplify your routine.

THE BATHROOM IS THE MOST WASTEFUL ROOM IN YOUR HOUSE. SECOND ONLY TO YOUR KITCHEN.

SHAMPOO AND CONDITIONER

If you wash your hair every single day, your hair will need soap daily. By gradually adding longer periods of time in between washes, your hair will require fewer washings over time.

Since I still use soap, I've found that the best way to be zero waste in this department is to find bar shampoo and conditioner that either come package-free or in paper/cardboard. I can find these products at my local health food store, but they are also online. To use, simply lather the bar in your hand, then transfer the suds into your hair and work them through starting at the scalp. I encourage you to do a little bit of research online before picking out your first bar to make sure it will work for you.

Many shops have bulk soap pumps that you can use to fill your own shampoo containers. There are even companies online that have closed the loop and offer shampoo and conditioner in reusable bottles that can be sent back to the company for refills. Genius!

BODY SOAP

Aside from saving the package waste, bar soap has more environmental benefits. According to Zurich's Institute of Environmental Engineering, scientists have found that liquid soap has a carbon footprint ten times that of bar soap—from the plastic involved in packaging to the chemicals that go into creating the liquid soap in the first place.

Body soap is a little easier to find in paper or cardboard since most stores have bar soap. And if you're trying to figure out what to do with the paper or cardboard that the soap comes wrapped in, compost it. To use bar soap, lather it in a washcloth or on a natural loofah sponge and scrub-a-dub-dub.

SHOWER SPONGE

Not only do shower sponges and poufs harbor bacteria, but they're also made of plastic and will ultimately end up in the trash since they are not recyclable. Cotton washcloths will last a lot longer and are compostable and recyclable after they're worn out. They also make great cleaning rags. You can also use a natural loofah for exfoliating. You can find them at your local farmer's markets, and if you have a garden, try growing some on your own. I can find loofah seeds at my local garden store. Loofahs are pretty finicky, so I recommend sticking to the instructions on the seed packet, using compostable pots when transplanting, and harvesting before the first frost of the season.

COTTON BALLS

Whether you use cotton balls for makeup removal, ointment application, or fingernail polish removal, reusable cloth scraps can take their place. Use an old worn-out T-shirt that has been ripped into smaller pieces or even one of those worn-out, old washcloths.

TISSUES

Ripped-up T-shirts or flannel shirts can easily take the place of tissues. To make your own, cut an old cotton or flannel T-shirt into scraps about the size of a standard tissue. Then stuff them into a jar or any container. Put them on your coffee table, bedside table, or in the bathroom and use as needed.

Add a few drops of lime, lemon, and peppermint essential oils to your jar of tissues. These oils will help break up mucus and relieve nasal congestion.

I've found that my nose doesn't get red and raw like it used to with disposable tissues because the fabric is softer.

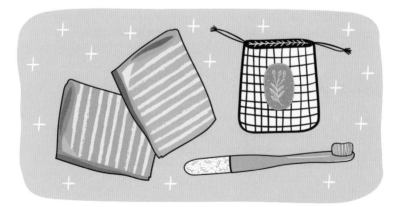

TOOTHBRUSH

Because of the size and type of plastics that make up toothbrushes, the likelihood of them getting recycled is slim. Sadly, this means just about every plastic toothbrush you've ever owned is still sitting or floating somewhere on this planet.

How can you keep your teeth and the planet clean at the same time? Try a sustainable alternative, such as a toothbrush made from natural materials like bamboo or one that has a reusable quality. Basically, avoid any toothbrush that would result in getting tossed into the trash can.

The "sustainable" toothbrush market is on the rise. Like any trend, there are companies that will jump on the bandwagon to make a quick profit without true values and sustainability in mind. To weed out any potential impostors, do your research before you buy and try to find companies that share your values when purchasing.

TOOTHPASTE

Toothpaste is an easy toiletry to make. With a few simple ingredients, you can whip up some toothpaste that's all natural, inexpensive, zero waste, and effective. Most people tend to think that fluoride is necessary in toothpaste to prevent cavities. Though it does help, you still get an adequate amount of fluoride in your drinking water. Also, your diet is important in fighting cavities and repairing tooth decay.

The ingredients to make homemade toothpaste can be found in the bulk dry goods section, bulk spice section, and in paper or cardboard packaging that can be composted later. I'm not a dentist (nor do I pretend to be!), so please do your research to find out if homemade toothpaste is best for you.

BILLIONS OF PLASTIC TOOTHBRUSHES ARE DISCARDED ANNUALLY, ACCOUNTING FOR MILLIONS OF POUNDS' WORTH OF WASTE IN OUR LANDFILLS.

HOMEMADE TOOTHPASTE

INGREDIENTS:

- ¼ cup (60g) coconut oil
- 2 tbsp (16g) baking soda
- 15 to 20 drops peppermint essential oil (optional)

INSTRUCTIONS:

1. Mix all the ingredients together and keep in an airtight container.

For most people, flossing is crucial to ensuring you don't have to keep going back to the dentist for expensive procedures. The problem is, flossing habits are not very sustainable.

Most floss is made of either nylon or Teflon, both of which are synthetically made plastics derived from extracted crude oil, a nonrenewable resource. Plastic doesn't break down quickly, and most of it will be around for longer than you will be alive. Instead of truly decomposing, most plastics photodegrade, which means that the sun itself causes the plastic to break down over a long period of time.

Luckily, there are many options to properly floss your teeth without contributing to the plastic waste issue. Here are a few:

Plastic-Free Floss: These plastic-free options are available and some even come in sustainable packaging. Look for brands that package their compostable floss in recyclable glass vials with metal lids. Some companies even make their floss containers refillable instead of requiring consumers to purchase new packaging every time.

Water Flosser: This device sprays water in a small, tight stream in between your teeth to wash away tiny food particles.

Compostable Thread: Have an old cotton or silk shirt that you no longer need? Unravel the threads and use them to floss your teeth! Once you're done with the thread, toss it into the compost.

Horsehair: Believe it or not, ancient people used horsehair to floss their teeth. The hair is rather coarse and very strong, making it a perfect substitute for plastic dental floss.

There are several ways to incorporate a zero waste flossing routine into your life. Just do a little research on each option and pick the one that's best for you.

EAR SWABS

Contrary to popular belief, ear swabs are bad to clean your ear canals with because swabs can push the wax deeper into your ear, risking infection or damage to the eardrum.

Your ears need to be cleaned, but not to the extent you may think. Wax is important in keeping the insides of your ear healthy. However, excess wax should be safely removed by using a warm washcloth to gently wipe the inside of your ear. If the wax is becoming a problem, have your physician remove it.

MENSTRUAL PRODUCTS

Menstruation costs us bleeding humans nearly $1,800 for just tampons in a lifetime. That's not including the pantyliners and other products that women also need.

Beyond the price, think about the environmental impact these products have. If you flush (please don't) or trash your disposables, they quickly add up and wreak havoc on the planet. And because they're made of plastic, pads and tampons will inevitably sit in landfills for a very long time.

There are also many health concerns associated with traditional menstrual products. Most of these products contain chemicals like dioxins, carcinogens, and reproductive toxins that have been linked to disruptive embryonic development, heart disease, and even cancer.

Since tampons are made of cotton, they more than likely contain the pesticides that are used in cultivating the crop. And it's the chlorine bleach that makes them white. Pads contain as many as four plastic bags' worth of plastic. Although companies claim these menstruation products are safe, there have been no conclusive studies to prove it. Check out these healthier alternatives to managing your menstrual flow.

MENSTRUAL CUPS

These cups have been around since the 1930s. A menstrual cup is a reusable, silicone cup that is inserted into the vagina. There are some major benefits to using a menstrual cup, including:

- It can stay in for up to twelve hours. You don't have to worry about changing it throughout the day like you do with tampons and pads.

- Unlike tampons, there's absolutely no risk of toxic shock syndrome.

- There's less waste. The average woman is expected to produce 62,415 pounds of garbage during her fertile years, which makes up about 0.5 percent of the yearly landfill waste. With the proper care, menstrual cups could last you up to ten years or even longer!

- You're not at risk of toxic chemical absorption.

- You'll save money!

Think about how much money you spend on period products now. You may spend $3 to $4 a month x 12 months = $36 to $48 per year x about 37 years' worth of cycles = $1,332 to $1,776. And that's just in tampons. Compare this cost to the roughly $120 worth of menstrual cups you'd buy in a lifetime if you purchased a new one every ten years.

CLOTH PADS

If cups aren't your thing, reusable cloth menstrual pads are great alternatives. They hold as much as a standard disposable pad, don't smell when changed at the appropriate times, and are very comfortable. To launder, place pads into the wash and use your regular detergent. If you are worried about staining, dry them out in the sun. And if you need to change while you're on the go, keep the dirty one in a wet bag (a bag that is lined and liquid proof) until you get home.

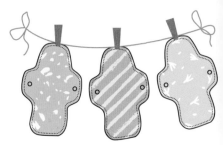

PERIOD UNDERWEAR

Period underwear has a built-in absorbing layer that can hold up to two tampons' worth of fluid. Depending on your flow, this comfy underwear can replace disposable pads and tampons. They can also be worn with your tampons or cup if needed.

APPLICATOR-LESS TAMPONS

Eliminating the plastic applicator from your tampon makes a difference because that plastic won't wind up in a landfill. The best thing about waste-free periods is that there are so many options that you're not limited to only one method.

RAZORS

The hardest part of shaving with a safety razor is getting over the fear of cutting yourself. You won't, at least not as much as you do with disposables. A safety razor is a reusable razor that uses replaceable, steel blades.

Let's talk about the environmental ramifications of disposable razors. According to the EPA, about two billion disposable razors are discarded every year.

Here are my tips on how to ease into shaving with a safety razor that will make you a pro in no time! Trust me, you'll never want to buy another disposable again.

Turn off the water. Because you want to take your time while you shave (and preserve water), I recommend plugging the drain while you wash your hair and using that water to shave your legs. I keep a cup or old plastic container in the shower to rinse off my legs.

Lather up. Rub some homemade shaving cream, shaving bar, or, my favorite, a conditioner bar on your legs.

Shave slowly. Shave in sections instead of pulling the blade all the way up your leg to prevent irritation.

Clean the razor. Once you're done, rinse the blade with water. Then, clean it off with a rag and rest it upright to dry. This will keep the blade in tip-top shape, prevent rusting, and extend the life of the blade dramatically.

Recycle your blades. Although the steel blade is made from recyclable material, recycling it can be a challenge. Blades are considered hazardous waste, so check with your local recycling facilities to be sure you can recycle them there. If not, take them to a local scrap metal facility.

But if you take care of your blades, they should last a long time.

Convince your friends to switch. Show your friends and family that shaving with a safety razor is not only more environmentally friendly, but it's also better for their wallets, easy, and probably the BEST shave they'll ever have. The next time they look at a disposable their noses will turn up!

If you're still not completely on board, that's fine. There are other options! Sugar waxing can be done in place of shaving, or maybe you want to stick it to stereotypical assumptions of how women should look and stop shaving completely. The choice is yours.

We use a lot of toilet paper. The demand is causing massive deforestation and several other detrimental effects.

If you want to go the no-TP route, keep a basket of clean cotton or flannel scraps to wipe after going to the bathroom. Throw used scraps into a container you keep next to the toilet and wash them once used. This is the same concept as cloth diapers and wipes.

Okay, I know you're wondering about the poop. To keep the gross factor at a zero, install a bidet, which is a toilet attachment that connects to your water line to clean your bum. Instead of wiping with toilet paper, you wash. I have a bidet,

and I will NEVER go back. You might wonder how a bidet could be any more environmentally friendly when it uses water. This is a valid point, but let's break down the resource usage of toilet paper compared to a bidet.

Making a roll of toilet paper uses 1½ pounds of wood, 37 gallons of water, and 1⅓ KWH of electricity. There's a lot that goes into producing toilet paper—and that's only the virgin toilet paper that makes up most the market.

On the other hand, a bidet that doesn't have to be continually replaced uses a fraction of the water used to make virgin toilet paper. In the end, bidets use considerably less water.

If toilet paper is the only option you want to stick with, no big deal! There are several brands that will ship toilet paper to you in paper packaging minus the plastic wrapped around it.

AMERICANS SPEND BILLIONS A YEAR ON TOILET PAPER, WHICH IS MORE THAN ANY NATION IN THE WORLD.

TOILETRY PACKAGING

Your bathroom is most likely full of face creams, lotions, toners, hair spray, mousse, hair pomade, shaving cream, and other products. There's so much we think we need to buy, but the real question is whether we need any of it in the first place.

These discarded items create a lot of trash, but to take it one step further, many of the beauty products you buy contain harmful ingredients that take a toll on your body and the environment. Because of this and my desire to reduce my trash to nearly zero, I decided to DIY most of my toiletries or use simple items that do the same job. Deodorant and lotion can be easily made with ingredients most of us have in our homes. See my recipes on pages 82 to 91.

HOMEMADE DEODORANT

INGREDIENTS:

- ¼ cup (30g) baking soda
- ¼ cup (30g) arrowroot powder or cornstarch
- 2½ tbsp (35g) coconut oil
- 15 to 20 drops of preferred essential oil (optional)
- 1 small container

INSTRUCTIONS:

1. In a small bowl, mix the baking soda, arrowroot, coconut oil, and essential oil, if using.

2. Transfer the mixture to a reusable container.

3. Use as much as you need to reduce body odor.

HOMEMADE LOTION/SHAVING CREAM

INGREDIENTS:

- ½ cup (120g) shea butter
- ½ cup (120g) coconut oil
- ¼ cup (60ml) aloe vera, gel or plant material (depends on your preference)
- 15 to 20 drops of your preferred essential oil (optional)

INSTRUCTIONS:

1. Add all ingredients, except the essential oil, to a double boiler, which is a bowl on top of a pot of simmering water that keeps the mixture from scorching from the direct heat of the stove.

2. Heat the ingredients, stirring until the mixture has completely melted and combined.

3. Strain the mixture to remove any lumps from the aloe vera plant (only if using plant material).

4. Add the essential oil, if using and mix together.

5. Put the mixture into the refrigerator until completely solidified.

6. Use a stand mixer or hand mixer to whip the shaving cream into white, fluffy peaks (about 2 to 3 minutes).

For shaving, I just use conditioner, but my lotion recipe also works great for shaving cream (see step 6). The shelf life of this shaving cream is about three to four months, but I assure you, you'll use it all before then.

Notes:

To make this low waste, look for the items package-free at a bulk store or your local farmer's market. Or get them in glass to recycle or upcycle later.

This shaving cream is meant to be incredibly moisturizing, so it may be slightly greasier than you are used to. Rinse your legs after use and wipe with a towel to remove excess.

The composition of the cream may react differently to where you live, depending on the climate.

Although the shaving cream is made of edible ingredients, I don't recommend eating it.

If you don't want to make your own, you could always fill up your own container with lotion from a package-free shop or just use coconut oil to moisturize.

TONER

I don't have a fancy recipe for toner because I only use either apple cider vinegar or whiskey. Yep, whiskey. While scouring through old Victorian books about beauty, I came across passages that refer to using whiskey to tone your face. And it works!

Most conventional products like hair products come packed with ingredients that could potentially be unsafe. Making these products yourself can reduce that waste and give a little control over what's in it.

DRY SHAMPOO

I found that when you continually strip your hair of its natural oils by washing it every day your scalp works ten times harder to replace them, which can cause a greasy hair problem.

HAIR PRODUCTS ARE NOTORIOUS FOR COMING IN PLASTIC. NONRECYCLABLE PACKAGING.

Training your scalp to work less may require you to wash your hair less. You will reduce water consumption by a lot using dry shampoo. I wash my hair every three days, and I use dry shampoo on the days I don't wash it.

This shampoo has only two ingredients! But because I have darker hair, I add cocoa powder to darken the shampoo, so I don't look like I have white powder in my hair all day.

HOMEMADE DRY SHAMPOO

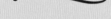

INGREDIENTS:

- 1 tbsp (8g) arrowroot powder
- 1 tbsp (8g) cocoa powder
- 1 small container

INSTRUCTIONS:

1. Combine the arrowroot and cocoa powders together until fully mixed. Transfer to a small container.

2. Apply to the roots of your hair using a powder brush.

3. Let the powder set for about 5 minutes, then massage it into your scalp.

4. Brush through with a comb, and you'll be all set to skip washing for another day!

Notes:
For darker brown to black hair, use more cocoa powder.

For blonde hair, omit the cocoa powder.

For auburn to red hair, try using cinnamon or paprika instead of cocoa powder, or mix the two to find your perfect shade.

HOMEMADE HAIR POMADE

INGREDIENTS:

- 2 tbsp (20g) grated beeswax
- 3 tbsp (42g) coconut oil
- 2 tsp (5g) arrowroot powder or cornstarch
- 1 small container

INSTRUCTIONS:

1. In a double boiler (see page 83), melt the beeswax.

2. Once the beeswax is melted, add the coconut oil.

3. Stir in the arrowroot powder and then pour into a container.

HOMEMADE HAIR SPRAY

INGREDIENTS:

- 1 cup (240ml) water
- 2 tbsp (25g) sugar
- 1 tbsp (15ml) vodka (use the cheap stuff)
- 1 small spray bottle (preferably glass)

INSTRUCTIONS:

1. Add the water to a pot and bring to a boil.

2. Add the sugar and stir until dissolved.

3. Add the vodka.

4. Let the mixture cool, then pour into the glass spray bottle.

5. Use as much or as little as you like.

Notes:

Get sugar in paper or in bulk.

Get vodka in glass and recycle later.

 **MAKEUP**

 **TONER**

I used to cake on my makeup. I used a ton of foundations, concealers, powders, bronzers, blushes, eyeliners, lipsticks, and a hundred different eye shadow colors. I would have sooner died than leave the house with no makeup. I didn't feel pretty in my own skin without needing to cover up what I thought was ugly.

Under all that makeup was another issue. My skin was terrible. I struggled with acne, excessive dryness and oiliness at the same time, and red patches galore. I needed to focus on taking better care of my skin, but makeup became my crutch. But I soon realized that I had to stop depending on makeup to cover my flaws and get to the root of the problem.

This started with establishing a nightly routine designed to let my face heal. Here I've compiled a few natural ways to take care of your face.

Whiskey is my go-to toner. I spray a little on with a small glass spray bottle. A cheap whiskey will do.

 **MOISTURIZER**

A good, natural moisturizer has been my secret weapon to healing my skin.

 **FACE WASH**

I wash my face with charcoal bar soap because it's gentle and gets off the daily grime. Using a washcloth to lightly scrub your face works to get any dead skin off as well.

I like to pamper the skin under my eyes a little more because the skin is much thinner and delicate. I use a bottle of vitamin E oil that I was able to get package-free from a bulk store I visited while traveling.

When my face finally healed, I found that I didn't need as much makeup. In fact, I no longer wear foundation—only a little blush that I make, some eye shadow, eyeliner, and mascara. It works, and I feel beautiful even without all the unnecessary junk I was slathering on my face. Because I use less makeup, I also save money and reduce waste by bringing less packaging into my home to begin with.

I buy several of my beauty items from responsible brands that package their products in reusable or recyclable materials. There are many companies that make, package, and ship makeup with the environment in mind. But it's also fun—and relatively easy—to give DIY a go. Let's start with blush.

To make homemade blush, look for beet powder in the bulk spice department. The work has been done for you and it costs next to nothing to fill a jar with it. If you don't have access to bulk beet powder, making it from scratch is easy.

Please note: This powder is a vibrant magenta color. If you want a deeper shade, you could leave it as is and simply apply it as your blush. Or you can use arrowroot powder or organic cornstarch to lighten the beet powder to get your desired shade. I get arrowroot powder from my grocery store's bulk spice area along with the beet powder. Organic cornstarch can be purchased in paper, which can be composted later.

HOMEMADE BLUSH

INGREDIENTS:

- 3 to 4 tbsp (18 to 24g) beet powder
- 1 to 3 tsp (3 to 9g) arrowroot powder or cornstarch
- Cocoa powder (optional, for another color variation, add as desired)
- 1 small container

INSTRUCTIONS:

1. If you have access to beet powder, mix the powder with arrowroot powder or organic cornstarch to get your desired shade. You can also add cocoa powder to make the color richer.

2. To make your own beet powder, first cut fresh beets into thin slices.

3. Place the beets in a dehydrator overnight or until completely crisp. Or roast in the oven on low.

4. Grind into a fine powder with a food processor, blender, or mortar and pestle.

5. Mix with arrowroot powder or organic cornstarch and/or cocoa powder for your desired shade.

Test some of my other tried-and-true makeup recipes for yourself on pages 90 and 91.

HOMEMADE LIP BALM

INGREDIENTS:

- 1 tbsp (10g) grated beeswax
- 1 tbsp (15g) coconut oil
- 1 to 3 drops of your preferred essential oil
- Reuse old lip balm container or small container

INSTRUCTIONS:

1. In a double boiler (see page 83), melt the beeswax.
2. Add the coconut oil and heat until melted.
3. Add a few drops of the essential oil of your choosing.
4. While still in a liquid state, transfer to the container.

HOMEMADE EYELINER AND MASCARA

INGREDIENTS:

- 3 charcoal capsules (I found them in a glass jar at my local pharmacy)
- ½ tsp aloe vera (I use my houseplants)
- ⅛ tsp vitamin E oil
- 2½ tbsp (37g) coconut oil
- 15 to 20 drops of essential oil (optional)
- 1 small container
- Mascara wand and eyeliner brush

INSTRUCTIONS:

1. In a small bowl, mix all the ingredients together. Transfer to the small container.
2. Apply normally with the mascara brush.
3. Apply with the eyeliner brush if using for eyeliner.

HOMEMADE EYESHADOW

INGREDIENTS:

- ½ tsp arrowroot powder
- ½ tsp bentonite clay
- Various powders for color choice (cocoa powder, ginger powder, spirulina powder, beet powder, turmeric, or charcoal)
- Vitamin E oil or aloe vera gel
- 1 small container

INSTRUCTIONS:

1. Mix the arrowroot powder and bentonite clay together in a small bowl.
2. Add your color choices—get creative! I usually mix various shades of charcoal powder, cocoa powder, and ginger.
3. Add as much oil or gel to make the mixture solid and non-dusty.
4. Pack it into a small container. Apply normally with an eyeshadow brush.

NATURAL MAKEUP REMOVER

INGREDIENT:

- Coconut oil

INSTRUCTIONS:

1. Apply coconut oil to a reusable cloth and wipe makeup off.

CHAPTER 5

SHOPPING

AND

WARDROBE

You'd probably be surprised to learn that the average home contains more than 300,000 items. It's no wonder rental storage units have popped up all over, encouraging you to house even more junk—most of which you'll never use.

⋘⋘⋘⋘⋘

According to the *New York Times*, rental storage has been the fastest growing segment of the commercial real estate industry over the past four decades. Sadly, people continue to acquire things with no intention of letting go.

I believe shopping is a way to cope with stress, anxiety, and depression. Shopping can be addictive because it gives you a sense of immediate accomplishment and instant gratification. These feelings of instant gratification and accomplishment increase our feelings of pleasure, or in this case, our levels of the hormone dopamine, which leads us to continue to shop. Usually, you don't buy what you need—you buy what makes you feel good.

Countless studies have shown that the more you own, the less happy you become. Now, I'm not telling you to purge everything from your closets. But I am suggesting that consumption can rule your life if it's not kept in check.

THE AVERAGE SIZE OF THE AMERICAN HOME HAS NEARLY TRIPLED IN SIZE OVER THE PAST FIFTY YEARS.

You will always NEED to buy things, but you can enforce ways to avoid impulse shopping or buying unnecessary items. Here are some helpful tips.

Use what you have. You don't always need to buy something new to do a job that could be done with something you already own. I find it's helpful to give myself a mandatory "shop block." When I think I need something, I wait thirty days before I buy it to see if I truly *need* it. Nine times out of ten, I can do without it.

Borrow what you need. Borrowing is like an elevated version of getting something secondhand. Just make sure you return it—and in good condition. If you can't find something you need from someone you know, there are several community-wide apps that let you rent and borrow stuff from others.

Get in the habit of swapping. Bartering or swapping is a great way to get something you need, get rid of something you no longer use, and avoid spending money. You can trade your goods and/or services for another person's goods and/or services (think mowing your neighbor's lawn in return for his free-range chicken eggs).

Consider secondhand shopping. Another option is to buy what you need secondhand from a thrift store. Anything you could ever need has already been made and used by someone else who's looking to get rid of it.

Give DIY a try. Learning to make necessary items is very empowering, and you can use your newly acquired skills to make gifts or teach someone else how to save money.

Buy locally. When all other avenues have been explored, buy new—but try to buy local first. And if you're wondering whether it's more environmentally friendly to shop at a local store or purchase something online, think about the travel miles a product has to take to get to you.

> IF ONE IN A HUNDRED AMERICAN HOUSEHOLDS SHOPPED RESALE, WE COULD COLLECTIVELY SAVE OVER $1.6 BILLION AND 1.1 BILLION POUNDS OF CARBON DIOXIDE EMISSIONS EACH YEAR.

Sometimes buying online is the greener option. However, purchase from small, ethical businesses online that are sustainable. If these businesses are continually supported, it's more likely they'll be around for a while.

Greenwashing

Keep a keen eye out for greenwashing, which is when false and unsubstantiated claims are made regarding the environmental benefit of a product or service. Many sleazy companies will try to weasel their way into the market to make a quick buck on your desire to make better choices. Because this is happening so much, it's getting harder to sift through the companies that are truly trying to make a difference versus the ones that are only in it for the money.

Transparency

Most good companies are completely open about their impact. Check their websites to see if they have listed the ways they're making change and have highlighted all aspects of their processes. From manufacturing to delivery, companies with good moral compasses will take you on that journey with them. However, other companies that don't have good moral compasses probably have nothing to report, considering they might not be as ethical as they'd like you to believe.

Disposability

Any products claiming to be green but are still disposable would fall into my definition of greenwashing. As a consumer, you must be smart about the real implications of the products you buy.

CLOTHING

Clothing and textiles account for a massive amount of global waste. And fast fashion is horrid. Trends and clothing lines change *every few weeks*, pushing stores to constantly change out their inventories. This means that millions of pounds of clothing goes completely unused (stores purposefully destroy clothing so that people can't get it out of dumpsters). And since styles change so much, the average consumer discards about 80 pounds of clothing each year. To make matters worse, most thrift stores are inundated with donated clothing. Many of these stores won't accept cheap, fast fashion items because of the poor resale value and bad quality.

THE EPA ESTIMATES THAT TEXTILES ACCOUNT FOR 5 PERCENT OF LANDFILL SPACE.

I know that keeping a bulging closet isn't healthy, so I have a "one in" and "three out" rule: for every article of clothing I bring home, I donate three pieces that I no longer wear. Here are a few tips to keeping your closet as environmentally conscious as you are.

Buy less. Make a conscious effort to buy fewer pieces of clothing. If 20 percent of the items in your closet are going unused, then you should work to minimize more. Take some time to assess your overall number of items and start minimizing your possessions, keeping only the things that have a purpose. This will make you feel less stressed.

Unsubscribe to marketing emails and remove yourself from mailing lists. If you no longer get sale notifications and "store bucks" in the mail, you won't know what you're missing. Also unfollow social media influencers who entice you to buy.

Keep a wish list. A list of items keeps you on track and allows for self-accountability. The great thing about the list is that you'll find that items are more likely to fall off then get scratched off. Establish a time period (a few weeks or months) in which to keep the items on your wish list. This gives you time to evaluate if you need it in the first place.

Check secondhand sources first. You've probably gathered that getting things secondhand is the better choice, but that doesn't mean settling for any item just because it meets the description on your list. Look for items that will last and are made from quality materials. Secondhand doesn't mean cheap quality. Ideally, when hunting down any item, look for things made from natural fibers and avoid fast fashion brands that design clothing to fall apart.

Here are some places that you can find secondhand clothing:

- Local Thrift Stores
- Consignment Shops (usually have better-quality items)
- Clothing Swaps
- Social Media Marketplaces
- Garage Sales
- Estate Sales
- Online Thrift Stores
- eBay

THE KEY IS TO BUY WITH PURPOSE.

Strategically buy new. When purchasing something new, like underwear, be strategic. Consider who you are purchasing it from. Support companies who are ethical and practice fair trade as well as organizations that care about the environment. And don't forget to invest in items that will last.

APPLIANCES

Research the best, most eco-friendly brands. Support companies that stand behind their product and look for reviews of longevity. You can shop for used appliances that are in good condition (look on social media marketplaces or online sites). And check with stores that have a dent-and-scratch policy where they will sell you something with a few dents and dings that is in perfect working condition for a discounted price. Not only can you get the appliance you want for a steal, but you'll prevent it from being shipped off for recycling.

SMALL APPLIANCES

These get tricky because so many small appliances are designed as throw-aways. First, evaluate if you really need the appliance in the first place. If you do, check with friends and family to see if they have one to spare. You can also research the most sustainable brands and check your local secondhand stores or social media marketplaces for them.

FURNITURE

Most furniture today is cheaply made in factories and manufactured with the worst materials, one of which is heavily chemical-laden particleboard. Look for furniture at thrift stores, estate sales, rummage sales, online marketplaces, or even the side of the road that is made of solid wood and built to last. Consider supporting local crafters and get a custom piece for your home—or even make it yourself if you know how!

KITCHEN ITEMS

Plates, cups, cutlery, pots and pans, and utensils can easily be found at a local secondhand shop.

TOOLS

If you don't plan on remodeling your home or doing a lot of DIY projects, I'd recommend borrowing or renting tools when you need them. However, it's important to have a tool kit full of basic tools like screwdrivers, a hammer, pliers, a socket set, and a drill.

HOME ELECTRONICS

Plastics, metals, and precious materials are manufactured and mined for the latest phones, smartwatches, and Bluetooth headphones. And the truth is, most electronics are designed to become obsolete within years, maybe months. Sadly, recycling isn't the answer because it can't keep up with the influx of quickly discarded products. In fact, most of the electronics we produce end up in the trash. Here are some quick tips to keep your home electronics in check:

1. Use your device for as long as you can.
2. Buy refurbished items.
3. Sell or trade any unwanted devices.
4. Don't try to keep up with the Joneses.
5. Make sure your broken device gets recycled.
6. If you don't need it, don't buy it.

RECREATIONAL GEAR

Consider borrowing or renting gear. This decreases the demand for new items and saves you space and money. There are plenty of online resources for renting gear for any activity.

OTHER WAYS TO AVOID IMPULSE SHOPPING

Here are some other ways you can ensure that the recent demand for cheap goods doesn't harm the planet.

Take care of what you already own. Increasing the longevity of each possession will reduce the demand for new products, thus decreasing the resources used to make these new products. When you take care of your possessions, you help eliminate tons of items from the waste stream.

Check manuals and YouTube tutorials. Seek out the manuals that came with your item for tips and instructions on how to fix it. Sometimes manuals provide an online source for tutorials. If you can't find any help through the manufacturer's materials, YouTube houses a wealth of video knowledge.

Contact the manufacturer for physical support. If fixing the item is outside of your skill set, contact the manufacturer to see if they will repair it for you. Sometimes what's broken could be as simple as one proprietary part that you could easily get or have the company fix for you. Just make sure the company is repairing instead of replacing. Even if the item is outside of your warranty, it's still worth a try.

Support a local repair person. You might be able to find someone locally to repair your item for a fee. I've used a repair person to fix my fridge and dishwasher a few times for less than what it would cost to replace it.

Seek out community repair groups. Some cities and towns have "repair cafés," which are basically workshops filled with donated tools. Volunteers are there to help you fix your items. Check online to see if there's a repair café near you.

Keep your stuff clean. Sometimes ensuring the longevity of your things can be accomplished by keeping them clean. If you put thought and care into purchasing and maintaining your items, you'll wind up spending less money and help create less waste overall.

CHAPTER 6

CHILDREN

AND

PETS

Raising children is expensive—and so is preparing for their arrival.

-‹‹‹‹‹‹‹‹‹‹‹-

Did you know that maternity wear is a $2 billion industry? And this is partly due to the inflated price of them being considered "maternity" garb. However, it's possible to get through an entire pregnancy without purchasing proprietary maternity clothing. Here's how to keep your clothing within your zero waste values:

1. Utilize what you already have for as long as you can.
2. Buy secondhand or borrow.
3. Buy sustainable pieces that can transition with you to an after-baby body.

I found that I already had clothing that would "grow" with me: baggier T-shirts, cardigans, button-up shirts, dresses, and so on. True, I had fewer clothing options towards the end of my pregnancy, but by then I was wearing the same two dresses nearly every day for the last couple of weeks leading up to the birth, and no one even noticed (or at least no one dared to say anything about it).

ON AVERAGE, PREGNANT WOMEN SPEND A WHOPPING $500 ON MATERNITY WEAR.

When I was pregnant, dresses were my best friends, and I couldn't live without belly bands that wrapped around my unbuttoned pants to make it look like I was wearing a layered shirt. And they were much more affordable than a new pair of expensive maternity jeans! Here are some ways to avoid inflated prices on maternity clothes.

Buy secondhand or borrow. You can borrow from a friend or look for online thrift/consignment sites that sell secondhand maternity clothes.

Buy sustainable pieces that can transition with you to an "after-baby" body. Buy clothing that you can transition into once you begin losing weight after giving birth. I bought a pair of black yoga pants that I wore up until I gave birth as well as a couple of nursing bras, tank tops, and dresses from a sustainable, organic maternity site that sells "nursing" clothing that works pre- and post-baby, which was important to me since I planned on breastfeeding. If you have to buy new, opt to support a sustainable and ethical company that does not support poor working conditions and clothing waste.

Find secondhand furniture for the nursery. There were only three pieces of furniture in my son Oliver's nursery: a chair, a dresser, and a crib. Take my advice and look for bigger pieces secondhand. Not only is this a more sustainable option, but it will also save you a lot of money. Check out local secondhand shops and online resale marketplaces to find larger pieces like the crib. These finds will add a uniqueness to your baby's room.

If secondhand isn't an option, look for items that are made locally or, better yet, make a piece yourself to celebrate the impending arrival of your little one! Lastly, if you must buy new and non-local, support sustainable and eco-conscious companies.

BABY CLOTHING

You can still find super adorable clothes for your baby while shopping secondhand. Also, take advantage of what friends and family have stowed away from when their kids were babies. You could potentially get most of the clothing you need for your baby for little money, while preventing countless resources and textiles from entering the waste stream.

EVERYTHING ELSE

Anything else you might need for your baby, including toys, books, and other gadgets, can be found used. Of course, there will be some things you'll want to buy new, so just buy those items with the intent that you'll never need to replace them. Also take into consideration that many of the items that are marketed to parents are completely unnecessary. Here's a quick list:

1 **Baby Changing Table:** The floor, couch, bed, kitchen table, or any flat surface will do.

2 **Foam Changing Pad:** I used a folded-up towel that I could easily toss in the wash.

3 **Diaper Bag:** I guarantee you have some old backpacks lying around that you could use. Don't buy yet another bag to fill with lots of stuff.

4 **Baby Shoes:** Really? Babies don't walk! Socks or bare feet will do just fine.

5 **Bibs:** Kitchen towels work perfectly. They hang lower to cover the baby's legs and are great for wiping hands and face after eating.

6 Baby Tubs: Use some folded towels instead and lay your baby on them in the tub or the kitchen sink. It's a lot easier.

7 Diaper Genies: Yuck. Do you really want to stick your baby's poo diapers in a bin that stays in the house? Take dirty diapers outside.

8 Bottle Warmers: You can heat up a bowl of water and stick the bottle in that for warming.

CLOTH DIAPERING

Cloth diapering is a symbol of green parenting, and for good reason, too. According to the EPA, millions of disposable diapers get landfilled. Though they only account for about 2 percent of landfill space, it adds up—not to mention that soiled diapers are riddled with bacteria and viruses that inevitably wind up in our water.

I understand that buying cloth diapers is an up-front expense that may not be feasible for everyone. Besides, I don't think buying all new cloth diapers in lieu of disposables is the greenest solution. It's not. In fact, it could be worse. Many people think cotton is so green, but it actually has a lot of dirty little secrets.

Remember, cotton is a crop. And with many mass-farmed crops, it's riddled with pesticides that can contaminate your water and ultimately cause harm if ingested. In addition to the pesticides, cotton requires a lot of water to grow.

When comparing resources used, cotton may be worse—or just as bad—as plastic. The difference is, cotton products are intended for reuse. Here's my opinion: choose secondhand cloth. That takes the need for new resources out of the equation and extends the life of an existing item.

By using cloth diapers, you'll have a little more laundry than normal. But once you get into a routine, it will become second nature. For one child, I'd recommend twenty-five to thirty cloth diapers. This will be a small investment up-front, but you'll save money in the long run, considering the average baby uses nearly 4,000 disposable diapers.

HOW TO HANDLE CLOTH DIAPERS

1 I use my homemade detergent you can find on page 129. If you don't have the time to make your own detergent, look for an eco-friendly detergent in either compostable or recyclable packaging.

2 There are several options to store dirty diapers without the stench. I used an old cat litter bucket with a lid. You could also use a diaper pail or a wet bag. I wouldn't recommend letting dirty cloth diapers sit for more than a couple of days because the last thing you'd want is for them to mold.

3 For the solids, dunk the cloth diaper into the toilet or get a sprayer attachment for your toilet to spray the poo particles off and into the bowl. Change, spray, toss, wash.

4 For wash day, I used a cold/hot cycle with a scoop of my homemade detergent. I would remove the inserts, make sure there was no exposed

Velcro, and wash. Once clean, I'd hang them to dry and never used the dryer. The sun is a miracle worker when it comes to eliminating stains. Besides, elastic bands wear out faster in the dryer.

5 For times when you're on the go, a wet bag, a watertight bag that can fit into your diaper bag, is a great place to keep dirty diapers until you get home.

DISPOSABLE DIAPERS TAKE ABOUT FIVE HUNDRED YEARS TO DECOMPOSE.

HOMEMADE BABY WIPE SPRAY

INGREDIENTS:

- 2 cups (480ml) warm water
- 1 tsp coconut oil
- 1 tsp vitamin E oil
- 1 pump or spray bottle

INSTRUCTIONS:

1. In a measuring cup with a spout, combine the water, coconut oil, and vitamin E oil and transfer to a pump or spray bottle for easy application. You can also keep some spray in a small bottle for your diaper bag.

Another great way to reduce diapering waste is to use reusable wipes rather than a package of disposals. Despite what some of the packages say, these are NOT flushable. To make matters worse, some disposable wipes get recalled due to bacteria.

Cut up an old flannel shirt and store the remnants in a jar next to your wipe station. Use dry ones and moisten with a homemade wipe spray. This lets you spray as much or as little as you'd like on the cloth wipe.

Minimalism plays a very important role in overall waste reduction. The fewer things you buy, the less waste you create. A diaper bag should have everything you need to reduce waste without being overpacked. Here are a few must-haves for your diaper bag:

- Organic Burp Cloth
- Wet Bag
- Rubber Pacifier (opt for a natural rubber pacifier instead of silicone)
- Cloth Wipes (keep them dry and throw into wet bag once dirty)

- Homemade Wipe Spray
- Organic Breast Pads (if you breastfeed)
- Toys (buy from companies that make sustainable, plastic-free toys)
- Cloth Diapers (look online or join a local cloth diaper swap and shops for secondhand cloth diapers—you'll save yourself money!)
- Extra Baby Outfit (two words: blow out)

BABY FOOD

Keep in mind that every baby transitions to solid foods differently. Please check with your pediatrician before introducing solid foods. My son Oliver's first solid food was rice cereal. Baby food aisles are lined with packaged cereals that are expensive, but they're relatively easy to make yourself (see the recipe on page 108). All you need is brown rice that you can get from the bulk bins.

HOMEMADE BABY CEREAL

INGREDIENTS:

- 1 cup (165g) ground brown rice
- 1 cup (240ml) water

INSTRUCTIONS:

1. Grind the brown rice in a food processor until it's a fine powder. Measure ¼ (35g) of the powder and set aside. Reserve the remaining rice powder in an air tight container for another meal.

2. Bring the of water to a boil in a saucepan and add the rice powder, whisking constantly for about 6 to 10 minutes or until cooked. Then remove from the heat.

3. Add breast milk or formula to the desired consistency.

4. Let the cereal cool until lukewarm, then serve.

It wasn't long until I started exposing Oliver to anything and everything I could. My tactic was to excite his palate as much as I could right out of the gate. And I'm sure you've guessed by now that I didn't buy any premade baby food. All you need is fruits, veggies, and a food processor. You can get fancy with your food combinations, or just stick with one to two items per puree.

HOMEMADE BABY FOOD

You'll want to start with fresh produce that has been completely washed. Set up a pot of boiling water with a steamer basket or use a colander if you don't have a steamer basket. Steam the produce until soft (just steam the ones that are naturally harder and not soft). Once soft, puree the produce in a food processor. Use veggie stock, water, breast milk, or formula to thin. You can serve to your baby immediately or freeze for later use. I found it easy to make large batches of baby food at once and freeze it in an ice cube tray. Once completely frozen, I'd transfer the food cubes into a glass jar that I kept in the freezer until it was time to warm up and serve to Oliver.

Once your little one enters daycare or school, you'll need to think about kid-friendly foods that come without packaging. Some good package-free school lunch options include:

- Banana and Honey Sandwich
- Hummus and Veggies
- Mini Blueberry Pancakes
- Pasta Salad
- Veggie Fried Rice
- Bagels from the Bakery

For produce, look for plastic-free options. You can usually find fresh produce at your local grocery store or farmer's market. Skip the produce that's pre-chopped because it's usually more expensive from what I've seen. And chopping veggies at home won't take a whole lot of extra time. For example, instead of buying baby carrots, get regular size carrots and cut them into strips.

If you eat meat or dairy, steer clear of all the prepackaged meats and cheeses. You can take your own containers to the deli counter.

To reduce waste, you'll have to make some things from scratch. This may sound daunting, but with a little extra prep, it can be done. Designate a day for your prep work and try to buy food items in nonplastic packaging. Here's how to pack a zero waste lunch:

Skip the disposables. You can do without the paper bags, zipper bags, plastic wrap, aluminum foil, and plastic utensils. Look around your home for things you already have— used lunch boxes, cloth bags, cutlery, and containers work perfectly for school lunches.

Use a reusable lunch box or cloth bag. Don't forget to check your local secondhand stores for some good deals on vintage metal lunch boxes. They will last forever!

Use a reusable beverage container for your drink of choice. Most thermoses will keep your child's drinks hot or cold for most of the day.

Forgo plastic zipper bags. You can use reusable zipper bags or any reusable container you have on hand. Everything from existing plastic containers, glass, and metal food containers to cloth produce bags, beeswax wraps, and cloth zipper bags work to store your child's lunch. Ask your child to bring home used silverware in his or her lunch container.

According to the National Retail Foundation, parents *spend billions of dollars* in back-to-school purchases. So, it's important that when you shop for your child's school supplies that you think long term and not just for the current school year. Get things with the intention that they'll last a long time. Your back-to-school expenditures will decrease over time when you buy items that last longer than one season. Besides, cheaper items usually wear out and need to be replaced much sooner.

ZERO WASTE SCHOOL SUPPLIES

Look around your home. Chances are, many items you already have will be on your back-to-school supply lists. Also, check out local secondhand shops. Rather than buying new, giving an existing item a second life takes a new product out of the waste stream.

BACKPACK

Look for a backpack that was made to last from a company that stands by their products. Having your child's backpack wear out after only a year is a waste of money. Choose solid colors so that they will not fall out of fashion. Especially with kids, getting something that's in style and trendy might not be a sustainable option.

PENCILS

Simple #2 pencils are still my preferred writing instrument over mechanical. They're inexpensive, write well, and can be composted once they're too short to use.

PENS

If you need to get a pen for your student, I recommend a nice fountain pen. In my opinion, they're a lot easier to use than a standard pen.

PENCIL CASE

Get a secure case that will last. Have a metal container lying around the house? Perfect! Your child can her keep pens and pencils in there.

These crayon creations make perfect favors, classroom gifts, Halloween goodies, Easter basket/egg fillers, or stocking stuffers. The best part is, they're completely zero waste because old broken crayons get used up and don't come wrapped in plastic, throwaway packaging. Here's how to make DIY recycled crayons:

1. Remove the paper wrapper around each crayon. You can put all the papers in your compost. While you're doing this tedious part, go ahead and preheat your oven to 350°F (180°C) degrees.

2. Break up the crayons into smaller pieces if they aren't already. Then separate them out if you want to have a color scheme. Put your crayon pieces into the mold of your choice. You can also use a cupcake tin if you don't have a silicone mold.

3. Place the mold into the oven. Mine took about 4½ minutes to melt (melting time depends on the depth of your crayon piles).

4. Once the crayons have fully melted, remove them from the oven. Let cool completely (about 30 minutes) before removing them from the mold.

CRAYONS

They come in paper and cardboard that can be composted after use. When the crayons get too short and broken to use, melt them down in a silicone mold to make new recycled crayons. These are simple to make! You can put them in different shaped molds, and kids love watching them melt in the oven. You can make everything from rainbows to hearts in a variety of colors or you can stick to one color scheme.

For the crayons, I use the same silicone mold I use for everything from dishwasher tabs and bath bombs to lotion bars and small candles. It has a nonstick surface that helps the crayons slide out after they have cooled.

SCISSORS AND RULERS

I'm sure friends and family have old scissors and rulers lying around to give you. If you can't find them secondhand, look for plastic-free options. Metal scissors and rulers will last a lifetime.

ERASER

Those standard pink, rubber erasers work great. Get the ones in a cardboard box so you can compost it later.

FOLDERS AND NOTEBOOKS

Look for ones made from recycled content. If you can find regular ones at the thrift store, that's great, too. Does your child have notebooks from the prior school year that are still in good shape? Just rip out and compost the used pages and reuse the notebook!

GLUE STICKS

Use natural, nontoxic glue sticks if you can find them. If not, it's not the end of the world.

CALCULATORS

Try to get calculators secondhand. Buy any items your child's teachers request like tissues, hand wipes, and paper towels. Our schools are already underfunded, so there's no need to put extra stress on your child's teacher for the sake of waste reduction.

Beyond the items you buy—or refuse to buy—for your children, you can empower them to make a difference in the world, respect the planet, and care for and value their possessions. Here are a few ways:

Get your kids outside. Encourage them to put down their mobile devices and go play outside. Bonus points if they play with a sibling or friend! They'll get exercise, not to mention the resources they'll be saving when they're not "plugged in."

Get your kids involved in gardening. If they grow their food, they'll be more likely to eat it. And if your kids eat the fruits and veggies off their plates, they're also helping to eliminate food waste. Teaching them how to grow food doesn't have to be elaborate or time-consuming. If you have a garden, great! But a cup of dirt and a few seeds works just as well. Let them get dirty! Show your children that growing their own food allows them to be sustainable and independent.

Give kids their own reusables. This includes reusable water bottles, napkins, a lunch box, and cutlery. Let your kids be in control of their waste avoidance by making reusable items a part of their everyday lives.

Encourage them to pick up litter.
Whether you're at the park as a family or on an evening walk, make it a habit to pick up litter. This is something my family does every time we go for a walk. We even sell the aluminum cans we collect to a local facility that gives us money by weight, and that money goes right into Oliver's allowance fund.

Let them help in the kitchen.
This is so much more than just teaching your kids to cook. It's about making and baking up memories that will last a lifetime. Besides, when you give your kids the skills to cook for themselves, they will become less dependent on fast food and prepackaged items.

PETS

Pets are adored by their owners—so much so that many people consider them to be part of their family, a lot like children. Of course, with pet ownership comes waste. I have cats, dogs, and a bird, so trust me when I say that a good quarter of my monthly waste is pet related.

According to Gregory Okin, a geographer at the University of California, Los Angeles, all the cats and dogs in the U.S. consume about 25 percent of all the animal-based products. This is bad for the environment because meat consumption, factory farming, and massive deforestation have wreaked havoc on our planet.

Pets are great, and they're here to stay, seeing as they add joy and longevity to our lives. But you need to be armed with the knowledge and tools to make

environmentally friendly decisions when it comes to your pets.

ACQUIRE PETS RESPONSIBLY

Millions of pets reside in shelters all over the world, and sadly, many of them will never find their forever homes. The costs associated with housing so many stray animals forces many shelters to enact kill policies if the pets aren't adopted after a certain length of time.

Many pet stores continue to fill their cages with animals that come from commercialized mass breeding organizations that profit from the quantity of animals produced rather than the quality of life those animals have. Not all breeders are bad, but before you adopt a pet, make sure you know your source and consider a shelter first.

FOOD

Shop in bulk and bring your own container to fill it up. There are several pet stores now that sell pet food in bulk. I recommend looking for wholesale pet stores for the bulk items. Also, make sure the quality of the food is still to your standards. I always value nutrition over packaging. I know how difficult it is to find pet food in bulk, so do the best that you can. Remember, zero waste is not about perfection.

Make your own pet food. It's possible to make pet food from relatively package-free ingredients. You can buy the raw ingredients instead of purchasing kibble or canned food. If you want to go this route, take your own containers to the butcher to have them filled.

There are mixed reviews about the benefits of making your own pet food. Do some research to see if this option is right for your pet. Keep in mind that there are additional vitamins and minerals that you'll need to buy separately and add to their food. Health always comes before waste.

Buy pet food in recyclable packaging. More and more pet food companies are "going green" when it comes to their food packaging. Go with this option if the food quality lines up with what you feed your pet. This isn't zero waste, but it's doing the best you can with the available options.

Reuse your nonrecyclable pet food bags. If none of the above suggestions works, you can always give your empty pet food bags a second life. You can fill them with thrift store donations, use them to transport mulch and soil home from the landscaping store, use them to store your compost, turn them into tote bags, or use them as garbage can liners.

PET WASTE

What to do with pet poop seems to be the biggest issue when greening a pet. Pet poop can contain toxoplasmosis, which is a disease spread through parasites that live in feces. Given the vast number of cats and dogs that populate living areas, pick up and dispose of your pet's poop right away, especially when you're in public parks or crowded areas.

LITTER BOXES

Keeping a zero waste litter box can be tricky, especially if you live in an apartment or rental. However, there are ways to do it—or at the very least, reduce the overall waste a litter box can produce.

Use sawdust, mulch, or dirt as your kitty "litter." Some cats will take to a new litter a lot quicker than others, so I recommend slowly incorporating the new stuff. You can get sawdust, mulch, and dirt package-free from your local landscaping store or even for free on certain sites. Most traditional pet litters come with fragrances to minimize odors. However, I just add baking soda to reduce the smell. Keeping the litter box clean also helps.

PET POOP BAGS

Poop bags are always an environmental cause for concern considering that most pet owners bag poop in plastic bags and then throw it away. I belief that your pet's health comes before waste reduction, so I suggest bagging the poop (just like every other pet owner does) and tossing it into a trash can.

There are still ways to make this option eco-friendly:

Use what you already have.
If you have plastic bags, zipper bags, or any other bags you're not using, save them as poop collectors. There's no sense in buying anything when you can get it right out of your own trash can. I'd even support grabbing plastic grocery bags out of the recycling bin at the store. The likelihood that those bags will ever be recycled is low anyway.

Collect usable bags while doing waste pickups. If I find usable plastic bags while picking up litter, I keep them to use for dog poop. Weird? Maybe. Effective? You bet.

Choose bags made with recycled content. Get bags made from recycled content, not biodegradable content. Biodegradable bags don't break down in a landfill and are the purest form of greenwashing.

TRAIN YOUR CAT

If you live in an area that isn't heavily trafficked, you could train your cat to potty outside just like you would train a dog. I would still recommend scooping and disposing of the poop, seeing as not having to mess with litter is still waste reduction.

BUY BULK LITTER

Some pet stores have bulk litter available. You get a bucket that can be refilled over and over, eliminating the need for new packaging. It may not be zero waste, but it does reduce the overall waste production. You can also purchase litter in paper bags or cardboard containers. This way, plastic waste is being reduced and the natural packaging can be composted.

COMPOST

If you use natural cat litter, you can compost the waste. And if you have a dog, composting can be an option as well. You just need to build a compost bin specifically for pet poop (keep it far from your garden). Most pet poop compost designs suggest housing it underground by digging a pit that's enclosed with a plastic trash can. Over time, the poop and litter in your compost will naturally break down and then you can use it—but I suggest using your compost for ornamental plants and not your garden produce just to be safe.

If you don't have your own yard to compost your pets' waste, check to see if your town's compost program accepts animal waste.

OUR PETS' ENVIRONMENTAL IMPACT GOES BEYOND THE WASTE THEY PRODUCE.

TOYS AND ACCESSORIES

Utilize items you already have or make your own. It's easy to make pet toys from old ripped-up clothing or by simply attaching a string to a stick. A cardboard box is a house favorite among my kitties that keeps them busy. Your pets don't need anything fancy if they have some sort of entertainment.

Pets can eat and drink out of bowls you already own, so there's no need to buy anything special. Also, check around your house for something you can repurpose into a litter box. It's always better to try and make do with what you already have than to buy new.

Buy all natural, quality items. Did you know that a lot of pet toys are made from or filled with plastic and are toxic? Regardless whether they're packaged or not, always opt for the all-natural, compostable toys. This ensures that when they wear out, they can be composted.

Seek out secondhand. Thrift stores are great for finding secondhand pet products such as toys, bowls, litter boxes, leashes, collars, you name it! Buying secondhand takes new products out of the waste stream and is better for your wallet. Also, check with friends and family for any items they can give you for free.

TREATS

Buy in bulk. Many pet stores have treats in bulk where you can skip the packaging and use your own containers. I sometimes fill a cloth bag with treats from my local pet store and then transfer them to a jar once I get home.

Make homemade treats. Consider making your pet treats from scratch. Many yummy treats can be made with relatively simple, package-free ingredients and are healthy for your pets.

ZERO WASTE DOG TREATS

INGREDIENTS:

- ½ cup (125g) peanut butter
- 1 banana, mashed
- 1 cup (120g) oat flour (if you can't find package-free oat flour, pulse rolled oats in a food processor until fine).

INSTRUCTIONS:

1. Preheat the oven to 350°F (180°C).

2. In a bowl, combine the peanut butter, banana, and oat flour until you form a dough ball.

3. Roll out the dough to about a ¼ inch (6mm) thick on a well-floured surface.

4. Cut the dough into shapes or use whatever cutter you have (even a pizza cutter will do).

5. Bake for 10 to 15 minutes, until crispy.

CHAPTER 7

HOUSEKEEPING

AND

PREVENTATIVE MAINTENANCE

Stores create visually appealing displays to push every cleaning product you'd never need. Even newspapers are stuffed with coupons for all types of cleaners.

⋘⋘⋘⋘

That's a lot of plastic-packaged products that will end up in the landfill. It baffles me that people don't question what's in household products. If it gets the job done, no one seems to care what's in it. Most of us tend to assume these products are safe.

Believe it or not, your entire home can be thoroughly cleaned with just soap and water—and the added help of vinegar and baking soda. I'm able to find all my cleaning products in nonplastic packaging. I get baking soda in cardboard, which I then compost. I buy vinegar in glass that I recycle or upcycle, and I can get soap package-free at a local shop.

I take the minimalist's approach with the cleaning tools I keep in my cleaning caddy. If you have a lot of items, use them up first. You're in a transition phase to zero waste, meaning that while you're using up your existing products, you will create waste, and that's okay. Don't throw out or give away products you already have for the sake of wanting to be "zero waste," because that's not very zero waste at all.

If you have the dusters with disposable sleeves or mops with disposable cleaning pads, use them up and then find a reusable alternative for the disposable part. Items that are designed to have something constantly replaced can be used with a reusable part instead. You don't have to give up the entire item. Wrap an old T-shirt around the mop head or sew a reusable dust sleeve for the dusting wand.

Here is the rundown on what I use to clean my home.

Cloth Rags

These are the ultimate cleaning powerhouse. They are the most versatile and helpful item. If you're not using paper towels made with 100 percent recycled content, then I hate to break it to you but trees are being cut down for you to clean up that mess of spilled juice. And let's not forget the gallons upon gallons of water that are used in manufacturing the paper towel, the fuel usage associated with hauling logs and finished product, the chemicals used in creating the paper towel, and on top of it, the fact that resources are continually gobbled up when packages are replaced. For heaven's sakes, just use a ripped-up towel or shirt. They are a whole lot less wasteful.

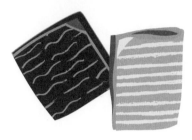

Plastic-Free Scrub Brush

Cloth rags don't have the scrubbing power like a good, plant-bristle brush. When it wears out, you can compost it. If you don't have a compost bin, bury it in the ground somewhere. Nature will do its thing.

Worn-Out Toothbrush

I can't even tell you how awesome of a cleaning tool an old toothbrush is. Those things will get up in all the crevices and hard-to-reach spots. They also do a great job at cleaning grout and getting in the nooks and crannies of small appliances or electronics. When you are at the time to switch over your plastic toothbrush to bamboo, don't toss it. Just stick it into your cleaning kit.

Old Newspaper

Rather than throw dreaded junk mail including newspapers into my recycling or compost, I put them to use. Other than wrapping gifts with them, I crumple them up for cleaning windows and mirrors. My grandmother passed this tip down to me.

Secondhand Cheese Shaker

Thrift stores are littered with these. I keep one for an easy way to sprinkle baking soda over surfaces.

Glass Spray Bottle

Upcycle a glass bottle or simply reuse an empty plastic spray bottle from cleaner. I use an old kombucha bottle with a spray nozzle that came off a bottle cleaner I used up. I keep my all-purpose cleaners in these. Remember to use what you already have.

Broom

I have a good sturdy broom that I hope lasts a very, very long time.

Mop

I do use a plastic mop—the one I had before I went zero waste. It takes a reusable cleaning head, and I refill the liquid compartment with my own homemade cleaner.

Bucket

Use any bucket you have lying around the house. I picked up a metal bucket from an antique store.

THE AVERAGE AMERICAN SPENDS $504 ANNUALLY ON PRODUCTS!

HOMEMADE ALL-PURPOSE CLEANER (SOAP)

INGREDIENTS:

- 2 cups (480ml) water
- 3 tbsp (45ml) liquid castile soap

INSTRUCTIONS:

1. Add the ingredients to a spray bottle and shake to combine.

2. Use a rag to scrub and wipe clean.

CLEANING THE KITCHEN

If I go a day without cleaning my kitchen in some way it usually means I'm not home. Dishes and grime pile up fast, especially if you have kids. And now that you are doing more cooking at home and the kitchen is getting a lot of use, there will be messes. Here's the thing: no one wants to spend their evening cleaning, so be strategic. Clean as you go or elicit some help from the troops (a.k.a. your family).

COUNTERS

I use one of two all-purpose cleaners I make for my home. I try to steer clear of using vinegar on porous surfaces, as the acidity of the vinegar could potentially damage them over time. If you have marble, granite, butcher block, or any other nonplastic counter surface, use the recipe on page 123.

KITCHEN SINK

I use baking soda and a scrub brush to get food grime off the side of my kitchen sink. I also wipe the water out of the inside after I'm done cleaning. Don't use baking soda if you have a porcelain sink because it could scratch it. Baking soda is abrasive, so it's best to stick with soap and water.

THE AVERAGE HOUSEHOLD CONTAINS 62 TOXIC CHEMICALS.

REFRIGERATOR

For nonporous things like appliances, I use my other citrus all-purpose cleaner that containers vinegar for a little added oomph. I take everything out of the fridge and spray down the shelves. I let the concentrate sit in heavily soiled areas and then wipe the surface clean with a reusable rag. For stubborn spots, I sprinkle on some baking soda and scrub. If the shelves are gross, I take them out and wash in the sink with warm, soapy water. After everything is clean, I leave a little bowl of baking soda in the fridge to keep it smelling fresh.

MICROWAVE

Like the fridge, I use my citrus all-purpose cleaner to clean the microwave. I remove the glass plate and wash with warm, soapy water. Then I spray the insides of the microwave and wipe clean with a reusable rag.

STOVE

I spray the top of the stove with my citrus all-purpose cleaner and use baking soda for any stuck-on grease or grime. For the oven, I make a paste to clean the inside. First, I sprinkle the oven with baking soda.

Then I spray it with vinegar. I let this paste bubble and sit for about 10 minutes before cleaning it off with a scrub brush.

If you want to add a little bit of shine to your stainless steel, rub it down with some olive oil.

HOMEMADE CITRUS ALL-PURPOSE CLEANER

INGREDIENTS:

- 1 cup (240ml) water
- 1 cup (240ml) vinegar
- Leftover lemon or lime peels

INSTRUCTIONS:

1. In a wide-mouth jar, add the water and vinegar.

2. For the lemon or lime peels, make sure as much of the flesh is removed, leaving only the skins. If any of the flesh is present, the solution will mold or ferment.

3. Add the peels to the solution and let sit for about a week.

4. Strain out the peels and pour the solution into a spray bottle.

Staying on top of cleaning the bathroom instead of neglecting this room is the best way to ensure that you don't have to spend an entire day cleaning it.

TOILET

I spray the outside of the toilet and wipe it clean with a reusable cloth rag. For the inside of the toilet, I use a homemade, fizzy toilet cleaner.

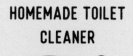

HOMEMADE TOILET CLEANER

INGREDIENTS:

- 1 cup (220g) citric acid
- 3 cups (540g) baking soda

INSTRUCTIONS:

1. Mix the citric acid and baking soda in a quart (1L) mason jar and store somewhere dry.

2. To clean the toilet, sprinkle a few tablespoons (40g) into the bowl, let fizz, then scrub.

3. Make sure you add the cleaner to your bowl only, not your tank.

MIRRORS

It's not necessary, but if you have a newspaper lying around, take advantage of its amazing, anti-streaking magic. I spray the mirrors with my citrus all-purpose cleaner and wipe clean with newspaper.

TUBS AND SHOWER

I'm not going to sugarcoat this and say, "Oh, my magical tub paste will make everything easy!" Because scrubbing a tub or shower just stinks no matter what. My tub and shower paste gets the job done, and that's the point.

HOMEMADE TUB AND SHOWER PASTE

INGREDIENTS:

- 1 cup (180g) baking soda
- ½ cup (120ml) liquid castile soap
- ¼ cup (60ml) water

INSTRUCTIONS:

1. Mix all the ingredients together in a bowl.

2. Smear the mixture on dirty areas or pour a little on a scrub brush and start scrubbing.

3. Rinse with water or vinegar.

Just like anything else, you need to keep a nondisposable mind-set—even with simple tasks like sweeping the floor. I have a vacuum that I don't use often. Instead, I try to keep my carpets and floors as clean as possible by not allowing shoes in the house.

SWEEPING

Sweep up the dirt, pick out any plastic nonsense, then fling the dirt outside. Don't put it into the trash.

MOPPING

I mop my floors with a very simple cleaner, just like everything else. I use lemony freshness in my home, so I add that to my basic soap cleaner.

DUSTING

I use my soap and water all-purpose cleaner (see the recipe on page 123), and a cloth rag. Dusting is fast, considering I don't have many knick-knacks. The less you own, the less you have to clean.

LAUNDRY

Have you ever read what's in the laundry detergent you use? There are sulfates, bleach, various dyes, fragrances, and many other hard-to-read ingredients listed on nearly every single bottle of detergent you'll find on store shelves. You may think that you need to use a complicated concoction of chemicals to get your clothes clean. That's simply not the case.

Making your own detergent is easy! All you need are three simple ingredients: washing soda, baking soda, and pure castile soap. When you make your own laundry powder, you'll also save a lot of money. Store-bought laundry detergent isn't cheap, unless you buy the cheapest bottle, which I can guarantee won't get your clothes clean. I used to spend about $15 a month on laundry detergent since the name brand jugs would last me about two months. Now, a month's worth of homemade soap costs me around $6 if I don't get any of the ingredients on sale.

There are several ways you can reduce your impact when it comes to the water and energy use your machines gobble up. It starts with how often you wash your things. By only washing items when you absolutely need it, you'll have less water waste and extend the life of your clothing.

Here's when to wash the following clothing:

Jeans: After five or six wears

T-shirts: After two or three wears

Sweaters: After two or five wears, depending on the material

Bras: After three or so wears

Underwear: After every wear

Skirts and dress pants: After five or seven wears

Sweatshirts: After seven wears

You can also save quite a bit of energy by washing your clothing in cold water. Only use heat when you really need to, like when washing whites or heavily soiled materials.

DRYING

I have a dryer, but I try to line-dry my clothing when I can. Anytime you can use less energy around your house is a win.

HOMEMADE LAUNDRY POWDER

INGREDIENTS:

- 1 cup (180g) baking soda
- 1 cup (180g) washing soda
- 5-ounce (140g) bar castile soap

INSTRUCTIONS:

1. Finely grate the bar of castile soap.

2. Mix the baking soda, washing soda, and castile soap together in a bowl.

3. Transfer the laundry powder into a container and keep by the washing machine.

4. Use about 2 to 3 tablespoons (24 to 36g) per load, depending on the size of the load and the soil level.

CHAPTER 8

HOLIDAYS, EVENTS, Parties, AND GIFTS

The holidays instill a sense of joy and wonder in an otherwise mundane world.

Sadly, though, companies prey on our love of the holidays by marketing everything we don't need to somehow elevate our holiday experience. Stores start preparing for the holidays months in advance. Holiday trees and decorations line store shelves beginning in July. So much marketing, preparation, and fuss to entice us to buy. And it's not hard to see—just look at the curbs during trash days littered with boxes from copious amounts of online purchases and abundance of packaging related to wrapping, food, and other unnecessary purchases, and more. During the holiday season, household waste increases by 25 percent.

Reducing waste during any holiday is completely manageable with a little forethought and determination. The trick is to assess each waste-causing activity and think of ways to improve and to purchase with purpose. Don't worry though, I've got you covered. I've broken everything down for you to help you avoid the waste, save money, and have a lot of fun in the process.

I've found that simplifying things during the holidays reduces a lot of stress. Did I mention that it saves a lot of money, too? Yes, I have, I know. Broken record. I've stopped obsessing over what I think is perfect and started enjoying what I have. It's also very helpful to let your creative juices flow and get crafty. Some of my favorite holiday traditions now involve making handmade items like drying fruit for the Christmas tree and collecting colorful fall leaves for Autumn window garlands. Before, I'd see something pretty online, then run to the store to purchase it. Now, I slow down and appreciate the beauty the seasons put before me.

People spend billions of dollars every year on Halloween costumes. And most costumes are not designed to last long—synthetic fabric and plastic masks aren't recyclable and made to be thrown away. Not to mention, they're not cheap! To reduce the expense and waste associated with costumes, you have a few options:

Use what you already have. Go through your closet to see what you can put together using clothes and items you already own. You could wear all black and draw whiskers with your eyebrow pencil to be a cat or wear a striped shirt, make a paper mask, and go as a burglar.

Ask a friend or relative if you can borrow a costume. Most people keep old costumes stashed away in the attic or basement and would be happy to lend one to you.

Go costume hunting at a thrift store. The thrift store is a gold mine when it comes to putting together a costume. Once Halloween is over, you can donate it back to the thrift store. Think of it as a rental costume!

Rent your costume. There are some cities that offer movie-style costumes for rent. If you're looking for something unique, this would be a fun and smart option.

Make your own costume. This is my go-to option. I make all my Halloween costumes because it's something I love to do. I usually strike gold at the thrift stores and don't have to buy anything new.

AMERICANS CONSUME NEARLY 600 MILLION POUNDS OF HALLOWEEN CANDY PER YEAR.

TREATS

Most treats are individually packaged. Imagine the amount of waste that produces.

I'm not going to tell you to prevent your kids from trick-or-treating or to turn your porch light off to avoid passing out candy. Halloween is my absolute FAVORITE holiday. With a little preparation and thought, you can add a little "green" to this black-and-orange holiday. Consider the following trick-or-treat alternatives:

Fruit

- Mandarin Oranges (can make them look like jack-o'-lanterns)
- Bananas (can make them look like ghosts)
- Apples
- Small Decorative Pumpkins

Natural Treats

- Gems
- Seashells
- Painted Stones

Useful Treats

- Boxes of Crayons
- Sidewalk Chalk
- Pencils
- Sprout Pencils (when they get too short to use, plant them in the ground and they grow into plants)

Homemade Treats

- Cookies
- Cupcakes
- Cake Pops
- Chocolates
- Caramel Apples

Treat Bags

- Bulk Candy
- Popped Popcorn

TIP

Write a fun note on the bags to remind kids to compost or recycle it after use. This treat is only zero waste if the recipient avoids tossing the bag into the garbage!

Seeds

- Packets of Herb Seeds
- Pumpkin Seeds

Games

- Bobbing for Apples
- Pumpkin Ring Toss
- Pumpkin Painting
- Halloween Scavenger Hunt

Palm Oil Free Candy

If you'd rather hand out packaged candy, opt for treats that are free of palm oil. A lot of the candy we consume is terribly destructive to endangered species' habitats. In fact, orangutans, Sumatran tigers, and Sumatran elephants have all been affected by deforestation due to palm oil farming.

Palm oil production and farming practices have also been found to violate many human rights and worker's rights. And because it's so cheap, it's used in almost half of the goods we Americans consume. Avoiding palm oil can be tricky because companies do NOT have to indicate "palm oil" on their ingredients' list. Palm oil can go by many names, so do research ahead of time to find candies that don't contain it.

THANKSGIVING

Oh, Thanksgiving! I dream of the bountiful smorgasbord of food all year long. My family doesn't make a lot of waste on Thanksgiving because we use real plates, glasses, and silverware. However, everyone's family gathering is different, so here are a few tips to help ensure that Thanksgiving waste is at the absolute minimum, whether you're hosting the holiday or a guest at the table.

Skip the foil and plastic wrap.

When transporting your side dishes, use reusable covers like beeswax wraps and silicone toppers. Or you can use a large towel to wrap your dish or just set a plate on top.

If you're hosting, use real plates, silverware, napkins, and glasses.

It's a fancy occasion—take out the real stuff! Don't worry about the dishes because family always steps in to help clean up.

Encourage guests to compost their scraps.

Hide the trash can and instead set out a bucket labeled "compost." You can even write a list of approved items that can be dumped.

As for items with animal products, those go in a separate container. Since animal products shouldn't go into your garden compost, bury them instead.

Opt for plant-based sides.
Encourage your guests to bring plant-based sides or prepare them yourself. Eliminating more items that contain animal products will help reduce the waste that mass factory farming produces.

Think twice about the turkey. Is turkey necessary? There are several other wonderful alternatives you could make on Thanksgiving like Tofurkey or veggie pot pie. If turkey is non-negotiable, consider buying a local, free-range, organic turkey.

Since so much food is served on turkey day, consider a few ways to dwindle food waste down to size. According to the USDA, the United States wastes nearly 40 percent of the food produced, annually. It's time to become a little more mindful of your holiday consumption, starting with food.

> WHEN IT COMES TO THANKSGIVING ALONE, AMERICANS WASTE ABOUT 200 MILLION POUNDS OF TURKEY.

IF YOU'RE HOSTING:

❶ Get an Accurate Head Count
Knowing exactly how many people will be in attendance lets you prepare the right amount of food. Without much excess, you can eliminate the need to figure out what to do with leftovers.

❷ Don't Encourage Everyone to Bring a Side
I know it's easy to ask everyone to bring a side dish to the holiday table, but that's just inviting too much food into your home. I only ask a few select people to help with sides and desserts.

❸ Keep It Simple
Come up with a menu and stick to it. The simpler, the better. When you keep the menu concise, you eliminate the need for dozens of ingredients that could potentially get tossed in the trash.

❹ Create a Leftover Station
Make it fun and easy for your guests to take food home by setting up a leftover station. Bring out the quart-size mason jars and fill them with leftover goodness for guests to take home with them. Leave a few empty ones out so guests can fill their own.

❺ Compost
Set out a bin for guests to scrape their food scraps into. Try to separate the meat out if you can, or if you're like me and have ample room for composting in your yard, just throw it all in.

IF YOU'RE A GUEST:

1 **Eat What You Serve Yourself**

I know it all looks (and smells) delicious, but trust me, the food isn't going anywhere. Pace yourself. Only serve yourself what you can eat so you have room for pie later.

2 **Bring Your Own To-Go Containers**

Leftovers from Thanksgiving are the best. Eating Thanksgiving leftovers prevents food from being wasted and saves you money because you won't need to buy groceries for a few of days.

Get creative with your Thanksgiving leftovers. There are hundreds of recipes online that will spark some inspiration and help you transform your stuffing and cranberry sauce.

THE HOLIDAYS

Around the holidays, most people feel compelled to buy and exchange gifts. Whether it's because exchanging holiday gifts has been a family tradition forever or because gift-giving is reinforced in movies, commercials, and on TV, this cycle of gift exchange comes to be expected.

Changing a tradition like this is very difficult to do. But if you've decided to forgo exchanging gifts, let people know your intentions well in advance. If you announce your anti-gift request early, it will plant the seeds necessary to start the transformation.

ANNOUNCE YOUR ANTI-GIFT REQUEST EARLY

You'll want to make it known that you don't want any gifts. This request isn't always well received, so be prepared to give people a list of things you would love to get in lieu of store-bought gifts, like cookies, candy, and experience gifts. To most people, at least this is the case in my family, when you say you don't want anything, they hear, "You don't have to get me anything," which means they'll still buy you something. I've found that it helps a lot when I'm very specific about what I'd like to get.

And talking about your new lifestyle of sustainable living A LOT will help. When you discuss our country's consumption and waste issues, people will listen—and you may eventually convince those around you that buying so much might not be the best thing to do. I've found that most people are genuinely interested in zero waste living and want to learn more about it, perhaps even jump on board.

GET YOUR FRIENDS AND FAMILY INVOLVED

Get them involved in your zero waste lifestyle. Making your friends and family feel like they are a part of this big picture will lessen your chances of getting "stuff" for the holidays because they won't want things either.

EXPRESS YOUR GRATITUDE

If you do get some gifts for the holidays, be thankful. This isn't the time to refuse or hurt anyone's feelings. The world could use more generosity and kindness, so show your appreciation.

And if you don't need the items you were gifted, you can return them for cash, regift them, or donate them to someone in need. But don't beat yourself up—changing people's mind-set about gifts isn't always possible. It takes a lot of work and patience. Another option is to gift items that are better for the planet, which will reduce the stress of finding the perfect gift and make you feel better overall.

On the following page, you'll see some zero waste gift ideas to consider.

ZERO WASTE GIFT IDEAS

1. Stainless Steel Straws
2. Audible Membership
3. Homemade Candy
4. Homemade Lotion
5. Plants
6. Concert Tickets
7. Crocheted Mittens, Scarf, or Blanket
8. Homemade Cookies
9. Vintage Jewelry
10. Candles
11. Homemade Salsa
12. Bamboo Travel Utensils
13. Restaurant Gift Certificate
14. Cotton Produce Bags
15. Beeswax Reusable Wraps
16. Massage Gift Certificate
17. Netflix Gift Card
18. Gym Membership
19. Fruit and Nut Basket
20. Homemade Bread
21. Dancing Lessons
22. Scuba Lessons
23. Package-Free Soap
24. Bath Bombs
25. Solid Shampoo
26. Cloth Napkins or Hankies
27. Homemade Scented Salts
28. Plastic-Free Scrub Brushes
29. Local Art
30. Reusable Bowl Covers
31. Homemade Lip Balm
32. Sugar Scrubs
33. Homemade Shaving Cream
34. Safety Razor
35. Stainless Steel Food Container
36. Bamboo Toothbrush
37. Homemade Perfume or Cologne
38. Cleaning Services
39. Cooking Lessons
40. Brewery or Winery Tour
41. Charity Donation in Recipient's Name
42. Homemade Hot Sauce
43. Digital Magazine Subscription
44. Ice Skating
45. Homemade Pet Treats
46. Locally Made Pottery
47. Guitar Lessons
48. Personalized Coupons
49. Movie Theater Tickets
50. Seeds
51. Compost Pail
52. Zip Lining

53. Wine or Craft Beer (make sure recipient recycles the bottles!)
54. Museum Tickets
55. Copper Pot Scrubbers
56. Homemade Spice Blends
57. Reusable Grocery Tote
58. Bulk Loose-Leaf Teas
59. Guided Tours
60. Homemade Pet Toys
61. Fountain Pen
62. Homemade Quilt
63. Secondhand Items
64. Gift Certificate for Family Photos
65. Tickets to a Ballet or Symphony
66. Handmade Wooden Toys
67. DIY Cookies in a Jar
68. Homemade Jams or Jellies
69. Homemade Soup Mix Kit
70. Lotion Bars
71. Organic Bamboo Cutting Board
72. Reusable Water Bottle
73. Reusable Lint Brush
74. Planet Love Life Bracelet
75. Bulk Coffee
76. French Press
77. Jar of Trail Mix or Nuts
78. Bulk Olive Oil and Balsamic Vinegar
79. Homemade Hot Cocoa in a Jar
80. Money
81. Un-Paper Towels
82. Homemade Vanilla
83. Infused Olive Oils
84. Reusable Coffee Cup
85. Handmade Socks
86. Organic Wool Dryer Balls
87. Reusable Cosmetic Rounds
88. Sporting Event Tickets
89. Wooden Bowl
90. Regift of Something You No Longer Want or Need
91. Reusable Sandwich Bags
92. Plant Trees in Honor of the Recipient
93. Homemade Candied Nuts
94. Homemade Playdough for Kiddos
95. Homemade Crayons for Kiddos
96. Bath Salts
97. Brownie Mix in a Jar
98. Holiday Light Tour
99. Tickets to a Dinner Theater
100. Bus or Train Credits
101. Quality Time Together

HOLIDAY WRAPPING

You spend hours meticulously wrapping each gift—wasting money on wrapping paper, disposable gift bags, and bows—only to have your loved ones destroy the gift wrapping, crumple it up, and toss it in a trash bag that's destined for the dump. Thankfully, there are alternatives to wrapping paper that are just as pretty and still do the job.

Furoshiki: This is the Japanese art of wrapping cloth. If you're like me, you have a cabinet full of tea towels, a bathroom full of bath towels, and a few sets of extra bed sheets. If you have babies, let's not forget all those swaddles and baby blankets. Use any of these cloth items to wrap your gifts! The recipient still "unwraps" the gift, but you get the wrapping back. If you want something more festive, buy holiday-themed cloth at a craft store and reuse it every year.

Cloth Gift Bags: Break out that sewing machine! All you need to do is fold a piece of fabric in half and sew the two sides closed, leaving an opening at the top. You can tie the open end with a ribbon for an easy, no-fuss gift bag you can use again and again. If you don't own a sewing machine or would rather not sew, you can wrap gifts with pillowcases or use reusable grocery bags for gifts. Remember, ribbon makes anything look more festive!

Clothing: You can use clothing like shirts to wrap your gifts. Simply slide the box inside the shirt and use the sleeves to tie it up. This could be a fun way for your family members to know which gifts belong to them. If it's wrapped in their shirt, it's their gift!

Reusable Tins: These are great for homemade goodies like cookies, candy, cakes, or peppermint bark. Of course, the contents don't have to be edible.

Upcycle: You can use just about anything—old newspapers, paper packaging, magazine pages—to wrap your gifts. Get creative!

Or don't wrap your gifts. Simple as that!

TREE DECORATIONS

In order to follow the zero waste principle, avoid buying anything new and put what you already have to use. You can string popcorn and cranberry garlands, make homemade ornaments, or make a wreath with some pruned evergreens. Find ways to bring the holidays into your home without the fancy-schmancy junk.

If you decide to have a tree, consider one that is in a pot and ready to plant after the season is over. Or you can get creative with what you already have by making one from cardboard or decorating a houseplant. Whether you go with a real tree or a fake one, make sure that you take ownership and discard of them properly when the trees are at the end of their lives.

REAL TREES

What do you do with a real tree once the season is over? Luckily for my family, we had a large plot of land with a pond that we'd sink the tree into for fish habitat. I know not everyone has his or her own fish pond, so here are some other ways to get rid of your real tree so it doesn't wind up in a landfill:

1 **City Tree Pickup/Drop-Off**
Most cities offer some kind of Christmas tree curbside pickup or drop-off. These services will take the trees, grind them into mulch, and use the mulch in city parks and flower beds. Contact your local parks and recreation department for available services in your area.

2 **Donate to the Department of Conservation**
Your state's department of conservation will most likely accept donated natural trees to aid the fish and wildlife habitats. Since many lakes and ponds are man-made, these bodies of water lack essential habitats for smaller fish and invertebrates.

3 **Donate to Your Local Zoo**
If you live near a zoo, call them and ask if they will take your tree for the animals. These trees provide hours of fun for the animals.

4 **Use It in the Garden**
The evergreen needles provide great nourishment for gardens. You can use the larger logs for flower bed barriers and compost the smaller branches.

FAKE TREES

While buying a fake Christmas tree made from plastic isn't the most eco-friendly choice, many people have already committed to one before forging on a new "green" adventure. Instead of throwing that tree out to sit in a landfill for a billion years, here are some ways to give your plastic needle conifer a new life:

❶ Donate to a Thrift Store or Nonprofit
Donate your fake tree to a thrift store or nonprofit organization and give someone else the opportunity to enjoy it around the holidays. I typically donate to the Disabled American Veterans organization.

❷ Donate to a Nursing Home
Most nursing homes and assisted living facilities love getting artificial trees. They use them as decorations in the lobby and give them to the residents to make their rooms festive.

❸ Donate to a Charity
There are many charities looking for artificial trees for families in need. This is a perfect way to celebrate the true meaning of the season by giving as well as letting that inner eco-warrior shine.

❹ Donate to a School
Many schools will graciously accept artificial trees for classroom decorating and theater productions.

The holidays don't have to be wasteful if you curb the temptation to over-consume and commit to letting go of things you no longer need or want.

EASTER

Like many religious holidays, Easter has been taken over by big corporations and their marketing tactics.

Today it's hard to discern exactly how much waste is generated during Easter, but from seeing how much you spend, you get the idea of how much of it is going straight to the landfill. The good news is, there are ways you can minimize your holiday waste and still enjoy Easter.

Look for alternatives to plastic eggs.
Plastic eggs will inevitably end up in a landfill due to their inability to be recycled over and over. It's best to look for alternatives, including:

- Fillable Wooden Eggs
- Fillable Cloth Eggs

Or skip the eggs and instead hide loose candy throughout the house. But if you'd rather have plastic fillable eggs and those are easier to obtain, look for them secondhand or ask friends and family if they have any to spare. I'm not suggesting that you get rid of your plastic eggs, but that you start to slowly phase them out.

Fill eggs with package-free goodies.
Easter eggs don't have to be filled with plastic-wrapped candy. You can fill them with waste-free items, including:

- Bulk Candy
- Dried Fruit
- Homemade Candy
- Money
- Homemade Playdough
- Clues to the Basket
- Secondhand Toys

- Pretty Stones or Gems
- Seashells
- Fun Coupons
- Chalk
- Trail Mix
- Seeds
- Crayons
- Homemade Lip Balm

Avoid plastic-wrapped, prepackaged Easter baskets.
You most likely already have a good vessel to hold those delicious Easter treats. Look around your house and use a basket, bucket, or even one of your child's toys as his or her Easter basket.

Instead of filling baskets with cheap plastic toys and candy, choose waste-free goodies (see the list of treats above), including:

- Homemade Baked Goods
- Package-Free Baked Goods
- Handmade Fabric Toys
- Sidewalk Chalk (in recyclable packaging)
- Secondhand Clothing
- Seeds
- Wooden Toys
- Experience Gifts

Skip the plastic Easter grass. This stuff is horrible for the environment. It can clog waterways, be ingested by pets and wildlife, and it's not recyclable.

Instead of using plastic grass, consider the following alternatives:

- Rip up paper from the recycle bins to fill the basket.
- Grow real grass in your basket or bucket.
- Put secondhand clothing gifts at the bottom.
- Use old hand towels as basket stuffing.

Use local eggs. If you decide to dye Easter eggs this year, make sure you buy them from local, free-range farms. Most of the white chicken eggs you buy from the supermarket come from massive factory farms that cram hens inside small cages to lay. The conditions are deplorable. You're probably thinking, "But aren't local eggs mostly brown?" Yes, but you can dye brown eggs.

Instead of using dye kits, consider using the following natural foods and spices to color your eggs:

- Red Cabbage: Blue
- Turmeric: Dark Yellow
- Saffron: Bright Yellow
- Spinach: Light Blue/Green
- Red Onion Skins: Dark Red

PARTIES/PICNICS/BBQS

The key to planning a waste-free party—or any get-together for that matter—is to think of all the stuff that generally makes parties so wasteful. Then just replace the waste-producing items with reusables. We've gotten accustomed to throwing elaborate, expensive parties with lots of balloons, streamers, themed disposable plates, napkins, cutlery, plastic party favors, mountains of gifts, and more. But parties don't have to be this way. Here are my tips for throwing trash-free, hassle-free parties that are still tons of fun.

No Disposables

I'm talking about the plates, cups, cutlery, napkins, etc. Don't waste your money on things that go into the trash. Instead, use real plates, glasses, cutlery, and cloth napkins.

It looks more elegant and is completely waste-free. If you don't have that many nondisposables, ask family and friends to help with additional silverware and extra plates instead of buying anything new.

Plant-Based Food

Reducing your meat and dairy intake has an enormous positive impact on the planet. Swap out the usual meaty sides or main dishes and add more plants to your spread. Corn on the cob, watermelon, spinach and strawberry salad, and veggie skewers are all yummy crowd-pleasers. Not everyone shares the same dietary preferences,

so you can even ask your guests to bring their favorite side dish. This will save you time and money.

BYOG (Bring Your Own Gear)

Encourage guests to bring their own plates and cutlery. Explain on the invitation that in order to make the event more environmentally conscious, you'd like to make the party a little less wasteful by avoiding disposables.

Compost Bin

Hide your large outdoor garbage bin and place a labeled compost bin next to the buffet tables. Put a spatula next to the bin to make it easy for guests to scrape uneaten food off their plates.

Dirty Dish Drop-Off

There are usually willing party attendees who will offer to help with the dishes. Aren't friends and family the best? If you don't have the extra help, set up a few bins with soapy water for guests to place their used dishes in. This will make it easier and faster to clean them all later.

Minimal Decorations

Skip the balloons, plastic table coverings, plastic party favors, and streamers. They're not necessary. You can keep the holidays waste-free by sticking to a few rules. Avoid everything you don't really need and focus on making memories.

CHAPTER 9

HOME MAINTENANCE

AND

Travel

The idea of zero waste isn't limited to your trash can.

~~~~~~~~~~

It's a concept that encompasses all the waste you and your family generate, especially wasted energy. When choosing to reduce your impact on the planet, sustainable lifestyle choices should go beyond the idea of the trash bin and composting. Here's what you can do to make your home more efficient.

Install a smart thermostat. Smart thermostats are electronic, programmable, self-learning, Wi-Fi-enabled devices that optimize the heating and cooling of your home to conserve energy. By using machine-learning algorithms, smart thermostats can build a custom heating and cooling schedule for your home based on temperature preferences and even shut off when you're not home.

Use smart outlets, plugs, and power strips. There are many electronic devices in your home that constantly use electricity, even when they're "turned off." This is because most devices like televisions, video game consoles, and media players remain in a state of "stand-by" in which it's waiting to be turned on. You could unplug each device after every use, but an easier solution is to plug these electronics into smart outlets, smart plugs, or smart power strips. These will completely shut power to the attached devices when you program them to do so. This saves energy and is an extra safety precaution for your home.

Become that plant person you always wanted to be. Many people use outdoor landscaping to reduce energy costs. For example, growing up in flat Missouri, people would plant cedar trees around the north sides of their property line as a defense against cold winter winds. This also helped reduce energy costs. Planting trees close to your home also helps shade your house when it's hot outside. The same is true for indoor plants. They not only help improve air quality, but they also help keep rooms cooler in the warmer months.

Use energy-efficient or smart bulbs. Consider replacing your old lighting with newer LED lighting, which uses less energy to illuminate the same amount of space. This is another way to reduce your environmental impact. Want to go one step further? Instead of using regular LED bulbs, purchase smart LED bulbs. These bulbs not only let you track usage and energy consumption, but they also let you control the light bulb itself. The most basic smart LED bulbs let you create light schedules or use your location to turn lighting on or off. More advanced smart LED bulbs have custom dimming output features, and some even let you make the light different colors.

Consider alternative energy sources. Solar panels or a wind turbine are greener ways to provide energy to your home. However, I understand that many people don't have this option, so another way to utilize green energy is to inquire what your current energy company provides. Some companies allow users to choose the source of their energy, whether it's solar or wind.

TIP

Don't forget to properly recycle your old bulbs. Never throw them in the trash! Some home improvement stores have recycling options for worn-out LED lights.

TRAVEL

When you're far from home, there are things you can do to reduce your impact on the environment. Whether I'm exploring new places or traveling for work, I do my best to be as sustainable as possible. It's easier to reduce waste when I'm at home, so being away adds another level of difficulty. But here are ways you can be more mindful about the environment, even when you travel.

Fly with sustainable airlines. Some airlines use biofuels on some of their flights to help reduce greenhouse gas emissions. Biofuels are derived from living matter like crops. Corn is a popular crop that's made into ethanol. Biofuels reduce our dependence on fossil fuels and reduce overall carbon emissions. However, it's still not a perfect solution.

Buy carbon offsets. A carbon offset is a reduction of carbon dioxide or greenhouse gas emissions used to compensate for or offset emissions produced from travel. In theory, it means that you would offset the exact amount of carbon damage you create by investing in a carbon "repair" somewhere else, like planting trees or engaging in carbon-reducing initiatives through environmental organizations.

Fly less often. Unfortunately, traveling is bad for the environment, and aircraft travel accounts for 12 percent of all U.S. transportation greenhouse gas (GHG) emissions and 3 percent of total U.S. GHG emissions. Try to limit your air travel to only necessary flights. Or if you must travel often, reduce your yearly flights by one or two per year.

Pack light. This lessens the overall weight of your luggage, thus reducing the amount of fuel needed to transport it. The heavier the cargo, the more fuel is used to transport it, which is why carry-ons are free and checked baggage costs more.

Use public transit. When you can, use public transportation or other means of low-impact transport. This not only reduces the overall emissions you produce, but also gives you another unique perspective as you explore new cites or countries. You may even meet locals who could give you recommendations for things to do or places to eat.

FOR SOME OF US, FLYING IS A MAJOR PORTION OF OUR CARBON FOOTPRINT.

Forgo the disposables. Travel with as many reusable items as you can. Swap out those plastic bottles of water for a reusable bottle, carry a reusable bag, and keep a few cloth hankies on hand. This will allow you to travel minimally, yet still have a few key items with you.

Use eTickets when you can. Going to a show or visiting a museum? Most organizations offer the option to "go digital" by offering eTickets.

Eat local when you travel. This option will have a lower impact than getting food that's been shipped from afar. Remember, your food has to travel to get to you, and a "food mile" is the distance a food item is transported from producer to consumer. Dine at local restaurants that offer farm-to-table options.

Look for local farmer's markets and bulk stores. Being in unfamiliar places makes it difficult to know where to find package-free options. My advice is to do your research beforehand. Also, ask your hotel's concierge. He or she may be able to direct you to the best places.

Bring your own package-free snacks and reusable bottle on the plane. You can refill your bottle with water once you get past security. Some flight attendants may be willing to fill your bottle for you, and most airports have water fountains in the terminals.

Choose sustainable toiletries. Get some reusable containers and fill them up with your toiletries from home. Or you can pack sustainable, package-free items.

Use a digital boarding pass. Trade your printed boarding pass for a digital copy on your phone. Many airlines have apps where you can keep flight details.

CHAPTER 10

THIRTY STEPS

TO

ZERO WASTE

CHALLENGE

We all want those around us to join our mission, especially when it's going to make the world a better place.

――――≪≪≪≪≪≪≪≪≪≪――――

And guess what? You can make a difference simply by sharing your passion for a sustainable lifestyle and talking about it. Express your reasons for doing it and tell others about how the changes have positively impacted your life.

Invite your friends and family to be a part of your mission. Ask them to join you on a trash pickup, go to a local Zero Waste support group, or even participate in an eco-friendly workshop. These are great ways to get others as excited about making a positive environmental impact as you are.

But don't expect those around you to jump on board instantly or even jump on board at all. Just push forward and be a beacon of environmentalism, because if you do, I can guarantee you'll influence a lot of people.

And lastly, never beat yourself up for making waste. It will happen; it's inevitable. Our economy is built on disposability and because of that, zero waste will always be impossible. It's not your fault. Focus on the things you can control and not the things you can't.

Now set forth and fulfill your waste-free destiny! Start here, with the Thirty-Step Zero Waste Challenge.

CHALLENGE #1

REFUSE EVERYTHING YOU DON'T NEED AND SEVERELY CUT BACK CONSUMPTION

Americans collectively use almost *two earths' worth* of resources, and that demand is on the rise. You must shift your mind-set on consumption, starting with reduction.

Your homework:

- Refuse what you do not need.
- Keep a list of things you want for thirty days before purchasing.
- Find ways to use what you already have instead of buying.

CHALLENGE #2

AVOID PLASTIC GROCERY BAGS

Plastic bags litter your roads, trees, parks, waterways, oceans, and life. According to the EPA, we use over 380 billion plastic bags and wraps yearly that require 12 million barrels of oil to create.

About 100,000 marine animals are killed each year just by plastic pollution, from sea turtles consuming plastic bags to whales washing ashore with over 60 pounds of plastic waste in their stomachs.

Your homework:

- Get some reusable bags secondhand or make some yourself.
- Keep your reusable bags on you to be prepared.
- Use up or recycle your existing plastic bag stash.

CHALLENGE #3

AVOID PLASTIC DISPOSABLE DRINK BOTTLES AND CUPS

Americans throw away roughly 50 billion (mostly) plastic bottles per year. And that's not even figuring in other types of disposable drinking vessels like coffee cups.

The Pacific Institute estimates that the equivalent of more than 17 million barrels of oil are used to manufacture the demands of plastic bottles each year—and it takes almost twice as much water to make a plastic bottle than what goes into the bottle after production. This process wastes nonrenewable resources and creates more virgin plastic that ends up in oceans, landfills, and parks.

Your homework:

- Find a reusable bottle or jar to use for "on-the-go" beverages.
- Look for plastic bottle litter and make sure it gets into the nearest recycling bin.

CHALLENGE #4

PHASE OUT PAPER TOWELS

This is a good opportunity to upcycle socks or any other piece of clothing that can't be saved. Cut the fabric into squares and then put them in a drawer or in a basket under the sink for easy access.

I can imagine that the idea of phasing out paper towels might be a challenge, but just keep an open mind and trust that a cloth rag can handle anything a paper towel can.

Your homework:

- Hide your current stash of paper towels for emergencies only.
- Find a few old, trashy shirts and cut them into rags.
- Keep rags in places you'd normally have paper towels.
- Use the rags the same way as you'd use paper towels.
- Wash cloth rags with towels.

CHALLENGE #5

GET YOUR "TO-GO" COFFEE IN A REUSABLE CUP AND REDUCE COFFEE CONSUMPTION

Americans drink 587 million cups of coffee per day. Instead of getting that coffee in a disposable cup, bring your own. Our coffee consumption is high and our environmental impact growing and harvesting coffee is very damaging. There's no reason we should be spending $5 per day on coffee. If anything, we should be saving money by making as much coffee at home as we can. Get beans in bulk, grind them yourself, and brew with a French press or use a reusable coffee sock.

Your homework:

- Get a reusable cup at home for coffee.

- Make more coffee at home.

- Reduce your overall coffee consumption. Drink tea instead if you need a caffeine boost.

CHALLENGE #6

STOP USING PLASTIC PRODUCE BAGS AT THE GROCERY STORE

A lot of produce naturally has its own packaging, so let your produce roll around naked in your cart. If you're worried about protecting your food from germs, remember that produce has already been exposed to many nasty things before even making it to the grocery store (wash it before you eat).

Cloth produce bags are smaller, reusable bags that can be used in place of the plastic sack. If your bags are a little heavier, write the weight onto the cloth bag and have the cashier take the difference off the total weight amount.

Your homework:

- Let your produce be naked.

- Get or make some cloth produce bags.

CHALLENGE #7

UTILIZE THE BULK BINS FOR DRY GOODS TO SKIP UNNECESSARY PACKAGING

More and more grocery stores are implementing bulk sections, which is fantastic for those of us who wish to eliminate food packaging. The trick? Bring your own containers. There are two ways to ensure you are not paying for the weight of your containers. Either utilize the tare option on the scale in the bulk aisle or write the weight of the container while it's empty, including the lid, onto the container for the cashier to deduct at checkout. Always check with customer service before doing this. Once you get your loose goods home, you can transfer them into jars that you can keep in your pantry.

Your homework:

- Locate the bulk stores in your community.

- Make a list of the items available in your bulk stores or bulk aisles.

- Shop bulk with your reusable containers to avoid packaging.

CHALLENGE #8

AVOID DISPOSABLE CUTLERY

It's estimated that 40 billion plastic utensils are thrown away every year in the United States, and that's usually after one use. Instead of using disposable cutlery, keep your own reusable utensils with you. Take a fork with you to work, stick one in your purse, or keep one in your car. Remember, it's all about using what you already have.

Your homework:

- Prepare yourself with your own cutlery.

CHALLENGE #9

REDUCE FOOD WASTE WHEN POSSIBLE

Look for ways to use up or preserve food so that nothing goes to waste. Challenge yourself to "eat your fridge" and clear out your pantries. Get creative with leftovers or simply freeze what you cannot finish. You can stretch your food and make sure nothing goes to waste in simple ways. Make pesto from the tops of carrots, turn overly ripe bananas into bread, or save your veggie scraps to make stocks.

Your homework:

- Eat all the food that you get.
- Freeze leftovers or serve them for lunches.
- Cook with scraps.
- Compost what you cannot eat.

CHALLENGE #10

SAY "NO" TO FREEBIES

When you accept freebies, you increase the demand for more to be made—and they take a ton of resources to produce. I'm not referring to getting something for free that you need. For instance, if you need a new pair of shoes and someone has a pair she'll give you for free, take it.

Your homework:

- Be confident and refuse the freebie if you don't need it.
- Say "no" politely and move on.

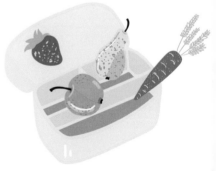

CHALLENGE #11

REPLACE WORN TOOTHBRUSHES WITH A BAMBOO ALTERNATIVE

Did you know about 50 million pounds of toothbrushes get sent to the landfill each year? Sadly, unless your toothbrush has been incinerated, it's still probably on earth somewhere. When your current plastic toothbrush is too worn out for teeth brushing, get yourself a sustainable toothbrush like one made from bamboo! They work just as well and can be composted when they wear out. If you don't have the ability to compost, simply remove the bristles from the brush since they are nylon (that will go into the trash, unfortunately) and push the bamboo handle into the ground to let nature do its thing.

Your homework:

- Use your current plastic toothbrush until it wears out.

- Add the plastic toothbrush to your cleaning supplies since they make awesome scrubbers.

- Find a sustainable, bamboo toothbrush company.

- Brush with bamboo!

CHALLENGE #12

TRY MAKING SOME OF YOUR TOILETRIES WITH SIMPLE INGREDIENTS

There are many bathroom items that can't be purchased in bulk, so making homemade toiletries is your next best bet and will save you money. If you don't have time to make your own toiletries, don't fret. You can support companies that choose sustainable packaging like cardboard, glass, or metal. See Chapter 4 for recipes.

Your homework:

- Use up what you currently have.

- Slowly incorporate sustainable products into your routine.

- Expect that the first recipe you try might not work for you, as our bodies are all different.

- Research sustainable and ethical companies.

CHALLENGE #13

PHASE OUT DISPOSABLE MENSTRUAL PRODUCTS FOR REUSABLE ALTERNATIVES

Periods are wasteful, stressful, uncomfortable, and expensive! A woman will use nearly 11,000 tampons or pads in her lifetime. That's about 62,415 pounds of garbage going into the landfill every year in the United States. On top of this, the products are made and soaked in chemicals that could be harmful to our bodies. Try menstrual cups, and if they aren't for you, there are many other options. Reusable cloth pads work well, and period underwear will do the job without the extra waste.

Your homework:

- Determine which option will be best for you.

- Find a sustainable and ethical company.

- Convince your friends to switch, too.

CHALLENGE #14

START COMPOSTING

Did you know that about two-thirds of household trash can be composted? Instead of throwing your biodegradable scraps into the garbage, start composting. You can start a compost pile in your yard or get a compost bin to allow the items to naturally decompose. If you have a garden, composting can give it a natural boost of nutrients and minerals.

Your homework:

- Figure out which composting option would be best for you.

- Save your scraps in the freezer if you don't have access to immediate composting.

- Have a family meeting on how to compost and where to put scraps.

CHALLENGE #15

SHOP YOUR LOCAL FARMER'S MARKET TO REDUCE YOUR FOOD'S OVERALL IMPACT

When you shop at your local farmer's market, you're investing money back into your local economies. Plus, you're out and about in your community, where you can meet local farmers, neighbors, and friends.

Buying produce that is grown locally shaves off a considerable amount of food miles, which equate to how far your food had to travel to get to your table. The more food miles, the more resources that went into getting that food to you. Plus, shopping at the farmer's market also means less packaging.

Food is at its peak deliciousness and nutrient level when it's grown locally. Produce that's in-season is also more abundant, making it a little more affordable. That means healthier, plant-based meals.

Your homework:

- Locate your farmer's market.
- Meal plan based on what's in season.

CHALLENGE #16

UNSUBSCRIBE FROM JUNK MAIL

It can take a lot of time and persistence to cut back your junk mail, but it's possible to get it under control. There are online removal services that will help you reduce your junk mail. But make sure you "opt out" of whenever you can because those services won't catch everything. Also be wary of contests, sweepstakes, or forms that require you to fill out your address—there's a good chance that if you do, you'll start getting junk mail soon after.

Your homework:

- Call companies or contact them via social media and ask to be removed from their mailing lists.
- Donate any unwanted magazines to a charity or doctor's office and recycle the rest.
- Look for a company name on promotional flyers and ask to be unsubscribed from their mailing list by contacting them via email, through online messaging, or by phone.
- Put a decal that says "No Junk Mail or Solicitations" on your mailbox. You can find these decals online.
- Get to know your mail delivery person to reinforce this.

CHALLENGE #17

TRADE THE PLASTIC–BOTTLED SOAPS FOR BARS

A lot of waste is generated from the soaps you use in the shower. Take advantage of bulk soaps, shampoos, and conditioners that are sold package-free or in paper. I can probably get three to four months out of one soap bar, and that's if I wash my hair every day. Money saved! You can find package-free soaps at farmer's markets and health food stores. You could also ditch the shampoo altogether by doing the "no poo" method, which means using baking soda, apple cider vinegar, or just water in place of shampoo.

Your homework:

- Use up the soaps you have first.

- Look for loose bar soaps at your local grocery and health food stores.

- Make homemade soap.

CHALLENGE #18

SAY "NO" TO DISPOSABLE STRAWS

Plastic straws may seem harmless, but we use and throw away 500 million of them a day! When going to a restaurant, it's probably inevitable that you're going to get a plastic straw in your drink. As soon as you order, politely ask the waiter not to include a straw.

Your homework:

- If you can go without a straw, do.

- Bring your own reusable straws when you dine out.

- Don't get discouraged because getting a plastic straw is sometimes unavoidable.

- Give the restaurant manager or corporate your opinion on their straw policy. Suggest that straws should be given by request only because of the detrimental impact they have on our planet.

- Tell your server ahead of time you do not want a straw.

CHALLENGE #19

GET SECONDHAND ITEMS WHEN NEEDED

Americans throw away 80 pounds of clothing a year. Discarded garbs account for 5 percent of what goes into landfills. It's important to try and buy pieces that are timeless, like a white button-down shirt. This will save you from constantly needing to purchase items that are on trend at the current moment.

Also, consider purchasing your clothing from secondhand sources. Buying secondhand not only takes clothing out of the waste cycle, but it also saves you hundreds—maybe thousands—of dollars! You can get other items like toys and books secondhand, and most everything will be package-free.

Your homework:

- Avoid buying new if you can.
- If you can't find anything secondhand immediately, keep searching.

CHALLENGE #20

BE MINDFUL OF HOW TRAVEL AFFECTS THE PLANET

Traveling, especially flying, has a giant carbon footprint. In fact, it's one of the most damaging things you can do to the planet. Yet the average number of frequent flyers is rising, the number of car owners is increasing, and public transportation doesn't seem to be getting better in some cities.

Your homework:

- Find ways to travel less frequently.
- Do a "no-car" challenge for a week. Walk or ride your bike places.
- Eliminate one flight per year.

CHALLENGE #21

GROW YOUR OWN FOOD TO CUT BACK ON WASTE AND FOOD MILES

Growing your own food is one of the most rewarding, sustainable ways you can get package-free produce into your kitchen. If you have a yard, putting in a garden is very easy. Fill up a spot or build raised garden beds, add some of that compost you've been working on, plant the seeds of your choice, and reap the benefits! You could also save seeds from your harvest to grow for next year.

And if you live in an apartment, you can still garden. Simply grow herbs on your windowsill or stick your celery "butts" and almost used green onions into water and watch them come back to life. Renting a plot from a local community garden is also a great option when you don't have the yard space. You can even grow food on the land of a friend or family member— and pay them back in free produce!

Your homework:

- Start something simple like a window herb garden.
- Find creative ways to use up every part of the plants that you grew.

CHALLENGE #22

PHASE OUT DISPOSABLE RAZORS FOR A REUSABLE ONE

According to the EPA, about 2 billion disposable razors are discarded every year. Yikes! The cost of safety razors varies depending on if you buy new or used. You can find secondhand razors at antique stores or even online. For the razor blades, you can get those at your local pharmacy. If you keep the blades dry after use, a box of them could last a few years.

Your homework:

- Save your disposables for travel.
- Research sustainable companies and get a reusable razor.

CHALLENGE #23

SIMPLIFY YOUR CLEANING PRODUCTS BY USING NATURAL CLEANERS

You can clean most of your house with simple ingredients like vinegar, baking soda, and water. To make cleaning zero waste, purchase vinegar in bulk, buy vinegar in glass jars to be recycled/ upcycled later, or make your own. For the baking soda, buy it in bulk or purchase it in paper packaging that can be composted. I mix up my cleaning solution and keep it in a glass spray bottle. You can also reuse any old spray bottles that you have lying around to prevent throwing them out.

Your homework:

- Use up the cleaners you already have and save the empty bottles.

- Use the empty bottles to hold your homemade cleaning products.

- Use cloth rags.

CHALLENGE #24

INCORPORATE MORE PLANTS AND LESS MEAT INTO YOUR MEALS

Livestock production has and is causing unbelievable amounts of damage to the environment through wasteful practices like the overuse of antibiotics and air pollution. Choose to participate in "Meatless Mondays" or make some meat-free lunches. Not only is consuming less meat beneficial to the environment, but it's also good for your health (not to mention your wallet!).

Your homework:

- Start slow: Pick one day of the week to go meatless.

- Make veggies the biggest portion on your plate.

CHALLENGE #25

REPLACE DISPOSABLE TISSUES WITH REUSABLES

Use a hankie or make your own homemade tissues by cutting up an old shirt or sheet and putting them into a container. Once the tissues or hankies have been used, throw them in with your normal wash.

There are some concerns about the sanitary situation of hankies while being sick, so I recommend using the handmade hankies for non-sick instances and the reusable tissues for when you're a germ factory.

Your homework:

- Use up what you already have and donate your extra boxes to a teacher.

- Use an old shirt or bed sheet to make some reusable tissues.

- Keep the tissues in the same places as you would the disposables.

CHALLENGE #26

REDUCE THE RESOURCE CONSUMPTION OF YOUR HOME

Being mindful of how to minimize excess usage in water and energy is good for the environment and for your wallet. There are many ways to make our homes more efficient, so after reading the following tips, I encourage you to dig a little deeper on your own.

Energy:

- Replace old incandescent bulbs with more efficient LED bulbs. They get 50,000 hours of illumination time and are safer to recycle since they don't contain mercury.

- Unplug appliances that are not in use. There are devices in your home that still use energy when turned off.

- When older appliances wear out, replace them with more energy-efficient counterparts— and don't forget to recycle older appliances.

- Change out your air filters regularly or upgrade to a reusable filter.

- Turn lights and fans off when not in the room. If you can, use solar panels.

Water:

- Add aerators with flow restrictors to your faucets to lessen the amount of water that goes through them.

- Add bricks or filled bottles to your toilet tanks to decrease the amount of water needed to fill it.

- Use a rain barrel to store water for your plants and garden.

- Set a timer for shower time.

- Keep a bowl in the sink to catch water to be used on plants or other areas.

- If it's yellow, let it mellow; if it's brown, flush it down. Sorry, I know that's gross.

- Utilize your compost pile rather than a garbage disposal because the disposal require a lot of water when in use.

- Install a water conservation system in your toilet.

- Check for leaks regularly.

Your homework:

- Work your way through the suggestions above.

- Research more ways to reduce energy and water waste in your home.

CHALLENGE #27

REPLACE DISPOSABLE PLASTIC WRAP WITH A REUSABLE ALTERNATIVE

There are a few alternatives to plastic wrap that are reusable and will save you money. Some easy solutions include putting a plate on top of a bowl instead of covering it with plastic wrap. There are also beeswax-soaked cotton cloths that cling over plates and bowls to keep in freshness. You can also use silicone bowl covers that create a seal around bowls and casserole dishes.

Your homework:

- Find alternative ways to replace plastic wrap.

FIX ITEMS WHEN THEY NEED REPAIR

Manufacturers have capitalized on our throwaway culture and have started mass-producing poor-quality items to satisfy our tendencies to toss out and buy new. This not only rewards plastic production, but it also puts more items into our waste stream.

In order to live more economically, repair and mend what you can. Learning to be resourceful and handy with keeping your items in tip-top shape saves a massive amount of waste and will also save you money. If you can't fix something yourself, get a list of reputable repair people who can help.

Your homework:

- Learn to sew and darn clothing that needs repairing (check out tutorials online). You may be able to send your clothes back to the manufacturer to be repaired.

- Take clothes you can't repair to a tailor and shoes to the cobbler to give these items a second life.

- For small appliances, locate small appliance technicians or call the manufacturer. If you call the manufacturer, they may send you the part you need for free.

- For larger appliances, check with the manufacturer to see if you are still within the warranty or if they can help.

CHALLENGE #29

EAT LESS FAST FOOD TO REDUCE OVERALL WASTE PRODUCTION

Every time I pick up trash in our neighborhood, I consistently find that 95 percent of the trash is from fast food, mostly cups. Not only is fast food unhealthy, but it also creates a ton of garbage! In one visit alone, the food comes wrapped in paper, shoved into a plastic bag or paper sack that is stuffed with paper napkins, plastic straws, plastic cutlery, and plastic-wrapped condiments. Did you know the biggest contributor to 49 percent of the waste that's floating in the oceans is from fast food? Instead of fast food, opt for wholesome meals with lots of fruits and veggies. Healthier food comes in less packaging, and you'll feel better too!

Eating In:

- You typically get less waste when you get your food to stay rather than at a drive-through.

- You also have more control over what you receive and what you can refuse when you're standing in front of the waiter or cashier.

Reusables:

- Opt for reusable alternatives—like your own utensils, cloth napkins, and reusable cups—instead of the disposable junk that comes with your meal.

Lids and Straws:

- These items are easy to refuse because they aren't really that necessary.

Real Glass:

- Many restaurants have real glassware behind the counter upon request.

Food To Go in Reusable Containers:

- Instead of getting your food to go in the disposable packaging, put it in your own containers. Call ahead to verify it's okay, then bring your containers to the counter upon pickup.

- Order your food to stay, then transfer it all to your containers yourself.

Restaurant Selections:

- Choose restaurants that have minimally packaged items, like sandwich shops.

- When you go to order, request that your food is minimally wrapped or not wrapped at all.
- Pizza is also a good option because it comes in recyclable and compostable cardboard boxes.

Your homework:

- Meal plan and meal prep.
- Choose restaurants that you can eat in and serve with reusables.
- Order items that have the least amount of packaging.
- Recycle or compost the paper and recycle any leftover plastic.

Recycling isn't the overall solution to the waste problem, but it's always better to recycle than discard items into the trash can. Depending on recycling to get us out of our trash problem is unrealistic. When it comes to plastic, most of it goes into a landfill even after it's been shipped to the recycling center, and sadly, most U.S. plastic waste is sold overseas. It's better to avoid plastic at all costs and recycle only metals and glass if you can.

Your homework:

- Start cleaning your recyclables. This increases their chance of getting recycled.
- Check with your local municipalities and recycling centers to see what they allow and don't allow. Follow the rules.
- Get a designated recycling bin and recycle everything you possibly can, but try to avoid the need to recycle in the first place.

RESOURCES

"Advancing Sustainable Materials Management: Facts and Figures." *EPA*. August 28, 2019. https://www.epa.gov/facts-and-figures-about-materials-waste-and-recycling/advancing-sustainable-materials-management-0

"Back-to-School and College Spending to Reach $82.8 Billion." NRF. July 12, 2018. https://nrf.com/media-center/press-releases/back-school-and-college-spending-reach-828-billion

"Bags by the Number." Waste Management. 2019. http://www.wmnorthwest.com/guidelines/plasticvspaper.htm

"The Be Straw Free Campaign." National Park Service. March 8, 2019. https://www.nps.gov/articles/straw-free.htm

"Before You Toss Food, Wait. Check It Out!" *USDA*. February 21, 2017. https://www.usda.gov/media/blog/2013/06/27/you-toss-food-wait-check-it-out

"The Best Way to Reduce Your Carbon Footprint Is One the Government Isn't Telling You About." AAAS. July 11, 2017. https://www.sciencemag.org/news/2017/07/best-way-reduce-your-carbon-footprint-one-government-isn-t-telling-you-about

"Bottled Water and Energy Fact Sheet." *Pacific Institute*. February 2007. https://pacinst.org/publication/bottled-water-and-energy-a-fact-sheet

"Coffee Grinds Fuel for the Nation." *USA Today*. April 09, 2013. https://www.usatoday.com/story/money/business/2013/04/09/coffee-mania/2069335

"The Coffee Industry Is Worse Than Ever for the Environment." *Huffington Post*. December 06, 2017. https://www.huffpost.com/entry/sustainable-coffee_n_5175192

"Comparing the Environmental Footprints of Home-Care and Personal-Hygiene Products: The Relevance of Different Life-Cycle Phases." October 21, 2009. Annette Koehler and Caroline Wildbolz. *ETH Zurich, Ecological Systems Design, Institute of Environmental Engineering. n.d.*

"Confronting Plastic Pollution One Bag at a Time." EPA Blog. November 01, 2016. https://blog.epa.gov/tag/plastic-bags

"Consumer Spending Statistics." Statistic Brain Research Institute. March 16, 2017. https://www.statisticbrain.com/what-consumers-spend-each-month

"Creative Ways to Cut Your Holiday Waste." EPA Blog. December 21, 2016. https://blog.epa.gov/2016/12/21/creative-ways-to-cut-your-holiday-waste

"The Downside of Halloween Candy."
American Institute for Cancer
Research. October 25, 2018.
https://www.aicr.org/press/press-
releases/2018/downside-of-
halloween-candy.html

"Eating Our Way Out of the
Plastic Waste Dilemma."
Plastics Today. April 15, 2016.
https://www.plasticstoday.com
/packaging/eating-our-way-out-plastic-
waste-dilemma/25470102124494

*The Environmental Consumer's
Handbook.* EPA. October 1990.
https://nepis.epa.gov/Exe
/ZyPDF.cgi/2000URC7.
PDF?Dockey=2000URC7.PDF

"Environmental Impact of
Disposable Diapers."
OurEverydayLife. December 11, 2018.
https://oureverydaylife.com
/environmental-impact-of-disposable-
diapers-5088905.html

"Exposure Assessment to Dioxins from
the use of Tampons and Diapers."
NCBI. January 01, 2002.
https://www.ncbi.nlm.nih.gov/pmc
/articles/PMC1240689

"Facts About Chlorine."
Centers for Disease Control and
Prevention. April 04, 2018.
https://emergency.cdc.gov/agent
/chlorine/basics/facts.asp.

"Food Recovery Hierarchy."
EPA. August 12, 2019.
https://www.epa.gov/sustainable-
management-food/food-recovery-hierarchy

"Food Waste Challenge."
USDA. July 23, 2019.
https://www.usda.gov/oce
/foodwaste/faqs.htm

"The Hidden Environmental
Costs of Dog and Cat Food."
Washington Post. August 04, 2017.
https://www.washingtonpost.com
/news/animalia/wp/2017/08/04
/the-hidden-environmental-costs-of-dog-
and-cat-food

"How Your Toothbrush Became
a Part of the Plastic Crisis."
National Geographic. June 14, 2019.
https://www.nationalgeographic.com
/environment/2019/06/story-of-
plastic-toothbrushes

"LED Facts."
Bulbs.com. 2019.
https://www.bulbs.com/learning/ledfaq.aspx

"Message in a Bottle."
Charles Fishman.
Fast Company Magazine.
July 2007: 110.

"Method Dish Soap, Basil."
EWG. January 10, 2013.
https://www.ewg.org/guides/
cleaners/5814-methodDishSoapBasil

"Niche Maternity Retailers Surge as Millennial Moms Redefine The Category."
Fast Company. October 05, 2016.
https://www.fastcompany.com
/3063624/niche-maternity-retailers-
surge-as-millennial-moms-redefine-
the-category

"Nondurable Goods:
Product — Specific Data."
EPA. November 06, 2019.
https://www.epa.gov/facts-and-
figures-about-materials-waste-
and-recycling/nondurable-goods-
product-specific-data

"Palm Oil — Deforestation for Everyday Products." Rainforest Rescue. 2019.
https://www.rainforest-rescue.org
/topics/palm-oil

"The Self-Storage Self."
New York Times. September 02, 2009.
https://www.nytimes.com/2009/09/06
/magazine/06self-storage-t.html

"Short History of Menstrual Cups."
Lunette. January 13, 2016.
https://store.lunette.com/blogs/news
/short-history-of-menstrual-cups

"Toxic Cleaning Products
That are Hurting Our Homes."
Molly's Suds. August 7, 2018.
https://mollyssuds.com/2018/08/07
/toxic-cleaning-products

"Why Secondhand Clothing Is Cool — For Your Wallet and the Planet."
Forbes. April 21, 2017.
https://www.forbes.com/sites/meimeifox
/2017/04/21/why-secondhand-
clothing-is-cool-for-your-wallet-and-the-
planet/#387f08bc33c3

"Why Toilet Paper Belongs to America."
CNN. July 8, 2009.
http://www.cnn.com/2009/LIVING
/wayoflife/07/07/mf.toilet.paper.history
/index.html

"You're Probably Going to Throw Away
81 Pounds of Clothing This Year."
Huffington Post. June 09, 2016.
https://www.huffpost.com/entry
/youre-likely-going-to-throw-away-
81-pounds-of-clothing-this-year_n_57572b
c8e4b08f74f6c069d3

INDEX

ABOUT THE AUTHOR

MEGEAN WELDON is a Midwestern girl who calls Kansas City her home. Since 2015, Megean has set forth on a journey to create zero waste. Known as The Zero Waste Nerd, she has re-evaluated her life to send nearly no trash to the landfill.

She started her blog Zero Waste Nerd **(zerowastenerd.com)** to show her progress of eliminating waste from day one. The transition made her more connected to food, it saved her a lot of money, and it made her healthier. She invites readers to join in on this journey.